55 Topics in Current Chemistry

Fortschritte der chemischen Forschung

Triplet States II

Springer-Verlag Berlin Heidelberg GmbH 1975

This series presents critical reviews of the present position and future trends in modern chemical research. It is addressed to all research and industrial chemists who wish to keep abreast of advances in their subject.

As a rule, contributions are specially commissioned. The editors and publishers will, however, always be pleased to receive suggestions and supplementary information. Papers are accepted for "Topics in Current Chemistry" in either German or English.

ISBN 978-3-662-15984-2   ISBN 978-3-540-37555-5 (eBook)
DOI 10.1007/978-3-540-37555-5

© by Springer-Verlag Berlin Heidelberg 1975
Originally published by Springer-Verlag Berlin Heidelberg New York in 1975
Softcover reprint of the hardcover 1st edition 1975

Library of Congress Cataloging in Publication Data. Main entry under title: Triplet states. (Topics in current chemistry; 54–55). Bibliography: p. Includes index. CONTENTS: 1. Devaquet, A. Quantum-mechanical calculations of the potential energy surfaces of triplet states. Ipaktschi, J., Dauben, W. G., and Lodder, G. Photochemistry of B, γ-unsaturated ketones. Maki, A. H. and Zuclich, J. A. Protein triplet states. [etc.] 1. Excited state chemistry–Addresses, essays, lectures. 2. Triplet state–Addresses, essays, lectures. I. Series.
QD1.F58 vol. 54–55 [QD461.5] 540'.8s [541'.24] 75–1466 ISBN 978-3-662-15984-2 (v. 1)

Contents

Springer-Verlag   D-6900 Heidelberg 1 · Postfach 105 280
                             Telephone (06221) 487-1 · Telex 04-61723

                             D-1000 Berlin 33 · Heidelberger Platz 3
                             Telephone (030) 822011 · Telex 01-83319

Springer-Verlag   New York, NY 10010 · 175, Fifth Avenue
New York Inc.   Telephone 673-2660

Characterization of Triplet States
by Optical Spectroscopy

Prof. Dr. Urs P. Wild

Physical Chemistry Laboratory, Federal Institute of Technology, CH - 8006 Zurich, Switzerland

Contents

1

I. Introduction

The primary photophysical processes occuring in a conjugated molecule can be represented most easily in the Jablonski diagram (Fig. 1). Absorption of a photon by the singlet state S_0 produces an excited singlet state S_n. In condensed media a very fast relaxation occurs and within several picoseconds the first excited singlet state S_1 is reached, having a thermal population of its vibrational levels. The radiative lifetime of S_1 is in the order of nanoseconds. Three main routes are open for deactivation:

First, fluorescence can be emitted;

secondly an internal conversion (IC) to a high vibrational level of S_0 and a subsequent vibrational relaxation may occur;

thirdly, an intersystem crossing process (ISC) to the triplet manifold, again followed by a fast vibrational relaxation, can form the triplet state T_1 having a thermal population of its vibrational levels.

The radiating transition from T_1 to the singlet ground state is spin-forbidden and the oscillator strength of the corresponding $T_1 \leftarrow S_0$ band is about 10^6 times weaker than that of an electronically allowed band. The radiative lifetime of T_1 is therefore quite large, of the order of 30 seconds for many hydrocarbons. The light-emitting process, which is called phosphorescence, competes with an intersystem crossing to a high vibrational level of S_0. This ISC route is quite often dominant and determines the observed lifetime of the triplet state in many conjugated systems.

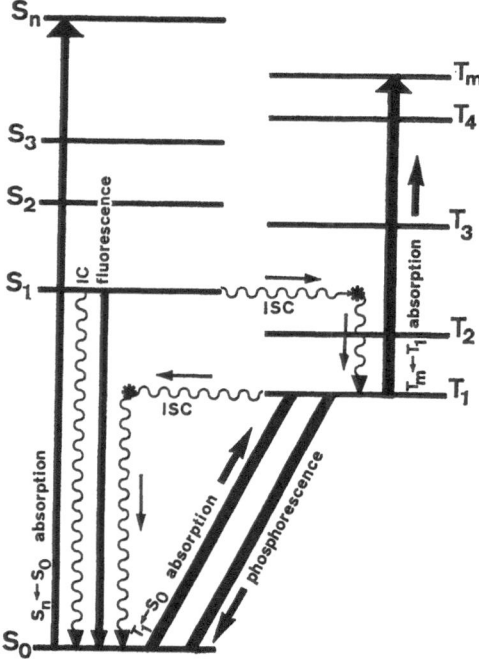

Fig. 1. Jablonski diagram. IC = internal conversion; ISC = intersystem crossing

Among the many excited singlet and triplet levels, S_1 and T_1 have distinct properties. They are in general the only levels from which luminescence is observed (Kasha rule); also most photochemical reactions occur from S_1 or T_1. Here we discuss the characterization of the lowest triplet state by electronic spectroscopy. First we treat the theoretical background that allows the absorption spectra of conjugated systems to be described, and then we discuss the routes that lead to phosphorescence emission and $T_1 \leftarrow S_0$ absorption intensity. Details of the experimental methods used to determine triplet–triplet and singlet–triplet absorption spectra, as well as phosphorescence emission spectra are given in Chapters III, IV, and V. Representative examples are discussed.

II. The Electronic Spectra of Conjugated Molecules

The semiempirical method of Pariser-Parr-Pople (PPP) [1-4] with configuration interaction is well suited to describe the electronic spectra of planar conjugated systems. The method is simple enough to treat relatively large molecules yet it still contains all the significant features necessary to explain the spectra of organic π systems. One of the main advantages for our purpose is that it is a many-electron theory and that it allows the construction of singlet and triplet wave functions, which lead to a basically correct description of the energy splitting between singlet and triplet states.

Even though a semiempirical method can never be used to "prove" or "disprove" an experimental "fact," it has a high stimulating power and allows the experimenter to classify and "understand" such facts.

1. Basic Assumptions of the Pariser-Parr-Pople Theory

The nuclear charges and the σ electrons form a fixed charge distribution, the so-called core, which is not affected by any change in the π-electron distribution. The singlet ground state wave function therefore describes only the π system and is given by a single closed-shell Slater determinant Δ_0 which is constructed from a set of π-molecular spin orbitals (SMO) $(\phi_1 \alpha)$, $(\phi_1 \beta)$, $(\phi_2 \alpha)$, etc.

$$\Delta_0(1\ldots n) = 1/\sqrt{n!} \begin{vmatrix} \phi_1(1)\ \alpha(1) & \phi_1(1)\ \beta(1) & \phi_2(1)\ \alpha(1) \cdots \phi_{n/2}(1)\ \beta(1) \\ \phi_1(2)\ \alpha(2) & \phi_1(2)\ \beta(2) & \phi_2(2)\ \alpha(2) \ldots \phi_{n/2}(2)\ \beta(2) \\ \cdot & \cdot & \cdot & \cdot \\ \cdot & \cdot & \cdot & \cdot \\ \cdot & \cdot & \cdot & \cdot \\ \phi_1(n)\ \alpha(n) & \phi_1(n)\ \beta(n) & \phi_2(n)\ \alpha(n) \ldots \phi_{n/2}(n)\ \beta(n) \end{vmatrix} \quad (1)$$

or in short notation

$$\Delta_0(1\ldots n) = 1/\sqrt{n!}\ |(\phi_1 \alpha)\ (\phi_1 \beta)\ (\phi_2 \alpha)\ldots(\phi_{n/2}\beta)| \quad (2)$$

Every occupied molecular orbital (MO) is taken with α and β spin and if there is symmetry degeneracy the whole degenerate set is included.

U. P. Wild

The MO's $\phi_j(i)$ are approximated by a linear combination (LCAO) of basis function $\chi_\mu(i)$ $(\mu = 1 \ldots B)$, which can be thought of as $2p_z$ functions centered at all the B atoms that contribute π electrons to the π system.

$$\phi_j(i) = \sum_{\mu=1}^{B} \chi_\mu(i) \, c_{\mu j} \tag{3}$$

It is assumed that the basic functions have zero differential overlap (ZDO):

$$\chi_\mu(i) \, \chi_\nu(i) = \delta_{\mu\nu} \, \chi_\mu^2(i) \tag{4}$$

The variational principle is employed to find the "best" set of molecular orbitals which can be represented by a LCAO and lead to a minimum value of the ground-state energy:

$$E_0 = \langle \Delta_0(1\ldots n) | H(1\ldots n) | \Delta_0(1\ldots n) \rangle_{1\ldots n} \tag{5}$$

where the spin-free Hamiltonian $H(1\ldots n)$ is given by

$$H(1\ldots n) = \sum_{i=1}^{n} H_{(i)}^{core} + \sum_{i<j}^{n} \sum^{n} \frac{e_o^2}{r_{ij}} \tag{6}$$

electron in the   electron–electron
field of the core   repulsion term

The variational principle leads to an effective one-electron problem and the "best" set of coefficients $c_{\mu j}$ $(\mu = 1 \ldots B)$ can be determined from [5]

$$\begin{pmatrix} F_{11} & \cdots & F_{1B} \\ \cdot & & \cdot \\ \cdot & & \cdot \\ \cdot & & \cdot \\ F_{B1} & \cdots & F_{BB} \end{pmatrix} \begin{pmatrix} c_{1j} \\ \cdot \\ \cdot \\ \cdot \\ c_{Bj} \end{pmatrix} = \varepsilon_j \begin{pmatrix} c_{1j} \\ \cdot \\ \cdot \\ \cdot \\ c_{Bj} \end{pmatrix} \tag{7}$$

where the elements of the "Fock" matrix $F_{\mu\nu}$ are given by:

$$F_{\mu\mu} = W_\mu + \sum_{\nu \neq \mu} (P_{\nu\nu} - Z_\nu) \gamma_{\mu\nu} + \tfrac{1}{2} P_{\mu\mu} \gamma_{\mu\mu} \tag{8}$$

$$F_{\mu\nu} = \beta_{\mu\nu} - \tfrac{1}{2} P_{\mu\nu} \gamma_{\mu\nu} \quad \mu \neq \nu \tag{9}$$

with the bond order

$$P_{\mu\nu} = \sum_{j=1}^{occ.} 2 \, c_{\mu j} \, c_{\nu j} \tag{10}$$

summed over all occupied MO's.

4

The valence state ionization potential $-W_\mu$, the resonance integrals $\beta_{\mu\nu}$, and the one-center electron repulsion integrals $\gamma_{\mu\mu}$ can be considered as basic parameters of the semiempirical method and can be adjusted to give optimal agreement. The core charges Z_μ indicate the number of π electrons the center μ contributes to the π system, and the two-center electron repulsion integrals $\gamma_{\mu\nu}$ are obtained from an empirical relationship such as the Mataga-Nishimoto formula [6]:

$$\gamma_{\mu\nu} = e_0^2 / \{ R_{\mu\nu} + 2\, e_0^2 / (\gamma_{\mu\mu} + \gamma_{\nu\nu}) \} \tag{11}$$

where $R_{\mu\nu}$ stands for the distance between the centers μ and ν.

The eigenvalues ε_j of (7) can be obtained from the roots of the characteristic polynomial of the secular equation:

$$|F_{\mu\nu} - \varepsilon_j\, \delta_{\mu\nu}| = 0 \tag{12}$$

and by inserting the eigenvalues in

$$\sum_{\nu=1}^{B} (F_{\mu\nu} - \varepsilon_j\, \delta_{\mu\nu})\, c_{\nu j} = 0 \qquad \mu = 1 \ldots B \tag{13}$$

the corresponding eigenvectors $c_{1j} \ldots c_{Bj}$ are determined. One major difficulty occurs in this eigenvalue problem. The matrix elements $F_{\mu\nu}$ can only be defined if the eigenvectors $c_{1j} \ldots c_{Bj}$ are already known! Still, solutions can be found by using an iterative procedure. First a reasonable set of zero-order coefficients $c_{\mu j}^{(0)}$ is assumed and used to calculate the $F^{(1)}$ matrix. The eigenvectors $c_{\mu j}^{(1)}$ of $F^{(1)}$ are then taken to calculate a $F^{(2)}$ matrix, and so on. As soon as the elements of the $F^{(i+1)}$ and $F^{(i)}$ matrix differ only by small amounts, self-consistency is reached, and from $F^{(i)}$ the self-consistent field orbitals SCF-MO's $\phi_j(i)$ are derived.

The lowest $n/2$ orbitals are doubly occupied and used to build the ground-state determinant wave function, while the rest of the orbitals, the virtual orbital set, will be used later to generate excited-state wave functions. The orbital energies and the SCF-MO's of naphthalene are given as an example and shown in Fig. 2.

2. Configuration Interaction

One might think of improving the description of the ground state by using not only one but a linear combination of determinants.

$$\Psi_0(1 \ldots n) = \sum_{s=0}^{p} \Delta_s(1 \ldots n)\, a_{s0} \tag{14}$$

The set of trial determinants Δ_s will of course include the closed-shell ground-state determinant:

$$\Delta_0(1 \ldots n) = 1/\sqrt{n!}\, |(\phi_1\alpha)(\phi_1\beta) \ldots (\phi_i\alpha)(\phi_i\beta) \ldots (\phi_k\alpha)(\phi_k\beta) \ldots (\phi_{n/2}\alpha)(\phi_{n/2}\beta)| \tag{15}$$

In addition, further determinants can be constructed by replacing for instance the occupied SMO $(\phi_i\alpha)$ in the ground-state determinant by a virtual SMO $(\phi_{i'}\alpha)$. Such a determinant is called singly excited with respect to Δ_0 and is given by

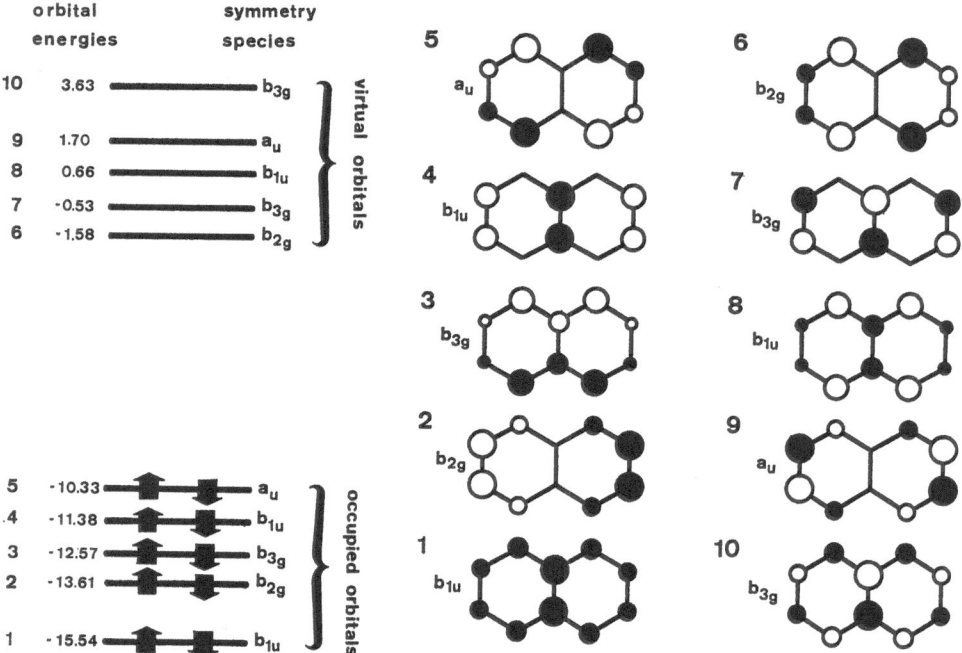

Fig. 2. Orbital energies (eV) and self-consistent field molecular orbitals of naphthalene. The molecule belongs to the symmetry group D_{2h}. The x axis is taken parallel to the long, the y axis parallel to the short molecular axis. All molecular orbitals are antisymmetric with respect to the molecular plane and belong to the a_u, b_{1u}, b_{2g}, and b_{3g} symmetry species. The areas of the filled and open circles are proportional to the square of the coefficients $c_{\mu j}$. The filled circles correspond to positive, the open circles to negative coefficients

$$\Delta^{i\alpha \to i'\alpha}(1\ldots n) = 1/\sqrt{n!}\,|(\phi_1\alpha)\,(\phi_1\beta)\ldots\underline{(\phi_{i'}\,\alpha)}\,(\phi_i\beta)\ldots(\phi_k\alpha)\,(\phi_k\beta)\ldots(\phi_{n/2}\alpha)(\phi_{n/2}\beta)|$$

$$(16)$$

Doubly excited determinants with respect to Δ_0 are obtained if two SMO's are replaced by virtual SMO's:

$$\Delta^{i\alpha \to i'\alpha}_{k\beta \to k'\alpha}(1\ldots n) = 1/\sqrt{n!}\,|(\phi_1\alpha)\,(\phi_1\beta)\ldots\underline{(\phi_{i'}\alpha)}\,(\phi_i\beta)\ldots(\phi_k\alpha)\,\underline{(\phi_{k'}\alpha)}\ldots(\phi_{n/2}\alpha)\,(\phi_{n/2}\beta)|$$

$$(17)$$

Higher excited determinants can be built in a similar way.

The determinants Δ_s form now a very convenient set of trial functions. The Raleigh-Ritz variational principle, keeping the determinants Δ_s fixed and varying only the coefficients $a_{1m}\ldots a_{pm}$, leads to the following matrix eigenvalue problem:

$$\begin{pmatrix} H_{11} & \cdots & H_{1p} \\ \cdot & & \cdot \\ \cdot & & \cdot \\ \cdot & & \cdot \\ H_{p1} & \cdots & H_{pp} \end{pmatrix} \begin{pmatrix} a_{1m} \\ \cdot \\ \cdot \\ \cdot \\ a_{pm} \end{pmatrix} = \begin{pmatrix} a_{1m} \\ \cdot \\ \cdot \\ \cdot \\ a_{pm} \end{pmatrix} E_m \qquad (18)$$

where the matrix elements are obtained from:

$$H_{st} = \langle \Delta_s(1 \dots n) \, |H(1 \dots n)| \, \Delta_t(1 \dots n) \rangle \tag{19}$$

If the number of determinants considered in (14) is increased, the lowest eigenvalue E_0 of (18) will be lowered more and more, and in the limit of an infinite complete set of determinants the true eigenvalue W_0 of the Hamilton operator $H(1 \dots n)$ will be reached assymptotically, the convergence being rather slow (Fig. 3).

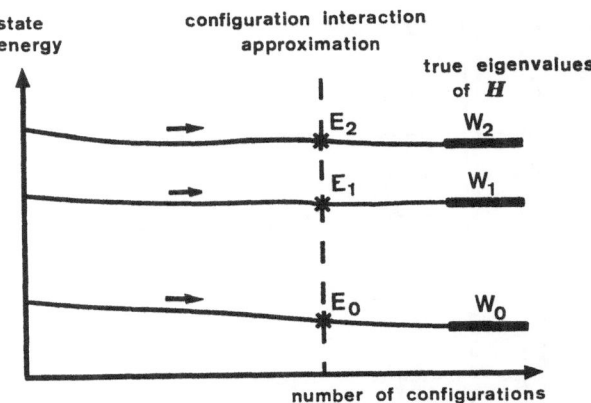

Fig. 3. Configuration interaction

The process of increasing the number of determinants just considered will also lead to a better description of the excited states. The eigenvalues $E_1, E_2, \dots$, obtained from a reasonably large configuration interaction problem, will represent acceptable approximations to the true eigenvalues $W_1, W_2, \dots$ of the H operator. Whereas it is known from the variational principle that E_0 is always larger than W_0, no corresponding statement can be made with respect to the excited energies E_1, E_2, etc. E_1 can be above or below W_1 and, if the size of configuration interaction is changed, even the relative position of E_1 and W_1 can be interchanged.

Basically, the configuration interaction procedure can be performed by using any orthogonal set of orbitals to construct the determinants $\Delta_s(1 \dots n)$. We shall now continue to show that the set of SCF-LCA-MO's determined in the last section is a particularly convenient choice.

If we choose only one determinant built from the lowest $n/2$ SCF-orbitals, the "configuration interaction" method will naturally give us $\Psi_0 = \Delta_0$ with the energy eigenvalue E_0 as the best ground-state description. This is clearly identical with the SCF result of the last section.

If, in addition, singly excited states with respect to Δ_0 are included, it can be shown (Brillouin theorem [7]) that the electronic ground state will still be described by the single closed-shell determinant wave function Δ_0 of energy E_0.

The wave function Δ_0 constructed from SCF orbitals is "so good" that it cannot be improved by the inclusion of *singly* excited determinants. The main effect of increasing the number of singly excited determinants in the CI problem will be a better description of the excited state levels.

Singlet and Triplet Configurations. Let us first consider an interaction which involves just the ground-state determinant Δ_0 and the four singly excited determinants $\Delta_1^{i\beta\to i'\beta}$, $\Delta_2^{i\beta\to i'\alpha}$, $\Delta_3^{i\alpha\to i'\beta}$, $\Delta_4^{i\alpha\to i'\alpha}$ which can be derived from an orbital promotion $i\to i'$ (Fig. 4).

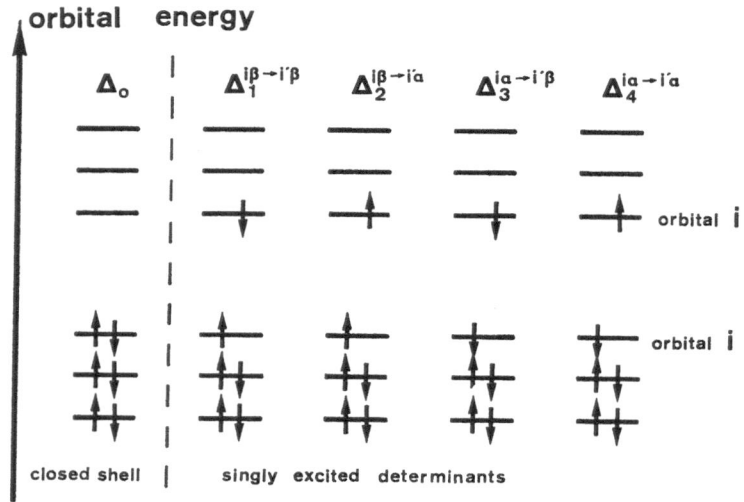

Fig. 4. Determinants used to derive singlet and triplet configurations

The matrix elements between the five determinants can be easily evaluated [12] and are given by:

$$H_{00} = E_0$$
$$H_{11} = H_{44} = E_0 + \Delta\varepsilon - J_{ii'} + K_{ii'}$$
$$H_{22} = H_{33} = E_0 + \Delta\varepsilon - J_{ii'}$$
$$H_{14} = H_{41} = - K_{ii'}$$

all other matrix elements are zero.

$\Delta\varepsilon = \varepsilon_{i'} - \varepsilon_i$ orbital energy difference

$J_{ii'} = \langle\phi_i(1)\,\phi_{i'}(2)\,|e_0^2/r_{12}|\,\phi_i(1)\,\phi_{i'}(2)\rangle_{1,2}$ Coulomb integral

$K_{ii'} = \langle\phi_1(1)\,\phi_{i'}(2)\,|e_0^2/r_{12}|\,\phi_{i'}(1)\,\phi_i(2)\rangle_{1,2}$ exchange integral

Setting up the matrix eigenvalue problem Eq. (18) for the five determinants gives:

$$\begin{pmatrix} H_{00} & 0 & 0 & 0 & 0 \\ 0 & H_{11} & 0 & 0 & H_{14} \\ 0 & 0 & H_{22} & 0 & 0 \\ 0 & 0 & 0 & H_{33} & 0 \\ 0 & H_{41} & 0 & 0 & H_{44} \end{pmatrix} \begin{pmatrix} a_{0s} \\ a_{1s} \\ a_{2s} \\ a_{3s} \\ a_{4s} \end{pmatrix} = \begin{pmatrix} a_{0s} \\ a_{1s} \\ a_{2s} \\ a_{3s} \\ a_{4s} \end{pmatrix} E_s \qquad (20)$$

The energies before and after the interaction are shown in Fig. 5. The eigenfunctions and eigenvalues are listed in Table 1.

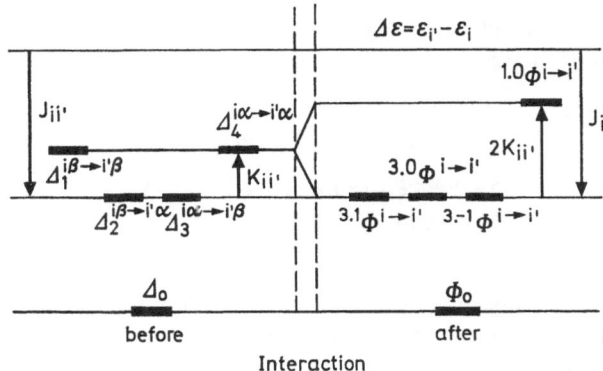

Fig. 5. Interaction problem involving the five determinants Δ_0, $\Delta_1^{i\beta \to i'\beta}$, $\Delta_2^{i\beta \to i'\alpha}$, $\Delta_3^{i\alpha \to i'\beta}$, and $\Delta^{i\alpha \to i'\alpha}$

Table 1. Eigenfunctions and eigenvalues of the interaction problem involving Δ_0 and the four determinants which can be formed from an $i \to i'$ orbital promotion

Eigenfunction	E	S	M_S	
$^{1,0}\Phi_0 = \Delta_0$	E_0	0	0	}Singlet S_0
$^{3,1}\Phi_1 = \Delta^{i\beta \to i'\alpha}$	$E_0 + \Delta\varepsilon - J_{ii'}$	1	1	
$^{3,0}\Phi_1 = \frac{1}{\sqrt{2}}(\Delta^{i\beta \to i'\beta} + \Delta^{i\alpha \to i'\alpha})$	$E_0 + \Delta\varepsilon - J_{ii'}$	1	0	}Triplet T_1
$^{3,-1}\Phi_1 = \Delta^{i\alpha \to i'\beta}$	$E_0 + \Delta\varepsilon - J_{ii'}$	1	−1	
$^{1,0}\Phi_1 = \frac{1}{\sqrt{2}}(\Delta^{i\beta \to i'\beta} - \Delta^{i\alpha \to i'\alpha})$	$E_0 + \Delta\varepsilon - J_{ii'} + 2K_{ii'}$	0	0	}Singlet S_1
	$\langle \Phi | H | \Phi \rangle = E$	$S^2\Phi = \hbar^2 S(S+1)\Phi$	$S_z\Phi = \hbar M_S\Phi$	

But this is not the full story. The Hamiltonian operator employed is a spin-free operator and does not work on the spin functions α and β. H commutes therefore with the spin operators S_z and S^2:

$$[H, S_z] = 0 \qquad [H, S^2] = 0 \qquad [S_z, S^2] = 0 \qquad (21)$$

9

with

$$s_x \alpha = \hbar/2 \; \beta \qquad s_x \beta = \hbar/2 \; \alpha$$
$$s_y \alpha = i\hbar/2 \; \beta \qquad s_y \beta = i\hbar/2 \; \alpha$$
$$s_z \alpha = \hbar/2 \; \alpha \qquad s_z \beta = -\; \hbar/2 \; \beta$$
$$S_z(1 \ldots n) = \sum_{i=1}^{n} s_z(i)$$
$$S^2(1 \ldots n) = S_x^2(1 \ldots n) + S_y^2(1 \ldots n) + S_z^2(1 \ldots n)$$

The commutator relations (21) show that the true eigenfunctions can be chosen such that they are simultaneously eigenfunctions of the operators H, S^2, and S_z:

$$H \; {}^{S,M_S}\Psi_n = E_n \, {}^{S,M_S}\Psi_n$$
$$S^2 \; {}^{S,M_S}\Psi_n = \hbar^2 \, S(S+1) \; {}^{S,M_S}\Psi_n \qquad (22)$$
$$S_z \; {}^{S,M_S}\Psi_n = \hbar \, M_s \; {}^{S,M_S}\Psi_n$$

It will prove advantageous to choose the approximate eigenfunctions of H — to be determined from the variational principle such — that they are true eigenfunctions of the S^2 and the S_z operators. If one works with the spin operators on the wave functions Φ given in Table 1, one immediately recognizes that they can be classified in a singlet ground-state function ${}^{1,0}\Phi_0$, in three degenerate triplet functions ${}^{3,1}\Phi_1$, ${}^{3,0}\Phi_1$, ${}^{3,-1}\Phi_1$, and in an excited singlet function ${}^{1,0}\Phi_1$.

For a better understanding of the energy level splitting of triplet and singlet levels T_1 and S_1, let us neglect the closed-shell electrons and consider just a two-electron system:

$$\begin{aligned}
{}^{1,0}\Phi_0 &= \frac{1}{\sqrt{2}} \; |(\phi_i \alpha)(\phi_i \beta)| \\
&= \phi_i(1) \, \phi_i(2) \cdot \frac{1}{\sqrt{2}} \, \{\alpha(1)\,\beta(2) - \beta(1)\,\alpha(2)\}
\end{aligned}$$

$$\begin{aligned}
{}^{3,1}\Phi_1 &= \frac{1}{\sqrt{2}} \; |(\phi_i \alpha)(\phi_{i'} \alpha)| \\
&= \frac{1}{\sqrt{2}} \, \{\phi_i(1)\,\phi_{i'}(2) - \phi_{i'}(1)\,\phi_i(2)\} \cdot \alpha(1)\,\alpha(2)
\end{aligned}$$

$$\begin{aligned}
{}^{3,0}\Phi_1 &= \frac{1}{\sqrt{2}} \left\{ \frac{1}{\sqrt{2}} \, |(\phi_i \alpha)(\phi_{i'} \beta)| + \frac{1}{\sqrt{2}} \, |(\phi_{i'} \beta)(\phi_i \alpha)| \right\} \qquad (23) \\
&= \frac{1}{\sqrt{2}} \, \{\phi_i(1)\,\phi_{i'}(2) - \phi_{i'}(1)\,\phi_i(2)\} \cdot \frac{1}{\sqrt{2}} \, \{\alpha(1)\,\beta(2) + \beta(1)\,\alpha(2)\}
\end{aligned}$$

$$^{3,-1}\Phi_1 = \frac{1}{\sqrt{2}} \, |(\phi_i\beta)\,(\phi_{i'}\beta)|$$

$$= \frac{1}{\sqrt{2}} \, \{\phi_i(1)\,\phi_{i'}(2) - \phi_{i'}(1)\,\phi_i(2)\} \cdot \beta(1)\,\beta(2)$$

$$^{1,0}\Phi_1 = \frac{1}{\sqrt{2}} \left\{ \frac{1}{\sqrt{2}} \, |(\phi_i\alpha)\,(\phi_{i'}\beta)| - \frac{1}{\sqrt{2}} \, |(\phi_{i'}\beta)\,(\phi_i\alpha)| \right\}$$

$$= \frac{1}{\sqrt{2}} \, \{\phi_i(1)\,\phi_{i'}(2) + \phi_{i'}(1)\,\phi_i(2)\} \cdot \frac{1}{\sqrt{2}} \, \{\alpha(1)\,\beta(2) - \beta(1)\,\alpha(2)\}$$

One sees immediately that the singlet functions have space parts that are symmetrical and spin parts that are antisymmetrical with respect to interchange of the electron coordinates, while the converse is true for the triplet functions. Since the Hamilton operator $H(1\ldots n)$ operates only on space parts which for $^{3,1}\Phi_1$, $^{3,0}\Phi_1$, and $^{3,-1}\Phi_1$ are identical, it is apparent that these three functions are energetically degenerate. The triplet functions $^{3,1}\Phi_1$, $^{3,0}\Phi_1$, $^{3,-1}\Phi_1$, and the singlet function $^{1,0}\Phi_1$ are constructed from the same set of orbitals and it is easy to show that they also have the same one-electron density $\varrho(\vec{r})$:

$$\varrho(\vec{r}) = \langle \Phi_1(1\ldots n)| \sum_{i=1}^{n} \delta(\vec{r}-\vec{r}_i)| \phi_1(1\ldots n)\rangle_{1\ldots n}$$

$$= |\phi_i(\vec{r})|^2 + |\phi_{i'}(\vec{r})|^2 \quad \text{for } \Phi_1 = {}^{3,1}\Phi_1, \, {}^{3,0}\Phi_1, \, {}^{3,-1}\Phi_1, \text{ and } {}^{1,0}\Phi_1. \tag{24}$$

They differ, however, in the pair density $\varrho(\vec{r}_1, \vec{r}_2)$:

$$\varrho(\vec{r}_1,\vec{r}_2) = \langle \Phi_1(1\ldots n)| 2 \sum_{i<j}^{n} \sum^{n} \delta(\vec{r}_1-\vec{r}_i)\,\delta(\vec{r}_2-\vec{r}_j)| \Phi_1(1\ldots n\rangle_{1\ldots n}$$

$$= |\phi_i(\vec{r}_1)\,\phi_{i'}(\vec{r}_2) - \phi_{i'}(\vec{r}_1)\,\phi_i(\vec{r}_2)|^2 \quad \text{for } \Phi={}^{3,1}\Phi_1, \, {}^{3,0}\Phi_1, \text{ and } {}^{3,-1}\Phi_1 \tag{25}$$

$$= |\phi_i(\vec{r}_1)\,\phi_{i'}(\vec{r}_2) + \phi_{i'}(\vec{r}_1)\,\phi_i(\vec{r}_2)|^2 \quad \text{for } \Phi={}^{1,0}\Phi_1$$

The pair density $\varrho(\vec{r}_1,\vec{r}_2)$ describes the probability that there will be simultaneously one electron at the position $\vec{r}_1$ and the other electron at $\vec{r}_2$. Corresponding pair densities for a few of the lowest singlet and triplet functions of an electron in a one-dimensional box are given in Fig. 6. The singlet functions S_0 and $S_2^{1^2 \to 2^2}$ are closed shells and have, of course, no corresponding triplet states. It is easy to see from the plots that all singlet functions have the following property: It is likely that we will find one electron at $\vec{r}=\vec{r}_1$ and the other at the same place $\vec{r}=\vec{r}_2$. For all triplet functions, the probability of finding two electrons at the same place $\vec{r}_1=\vec{r}_2=r$ is always zero. Since the average electron repulsion energy depends on the inverse of the average distance between the two electrons, the

11

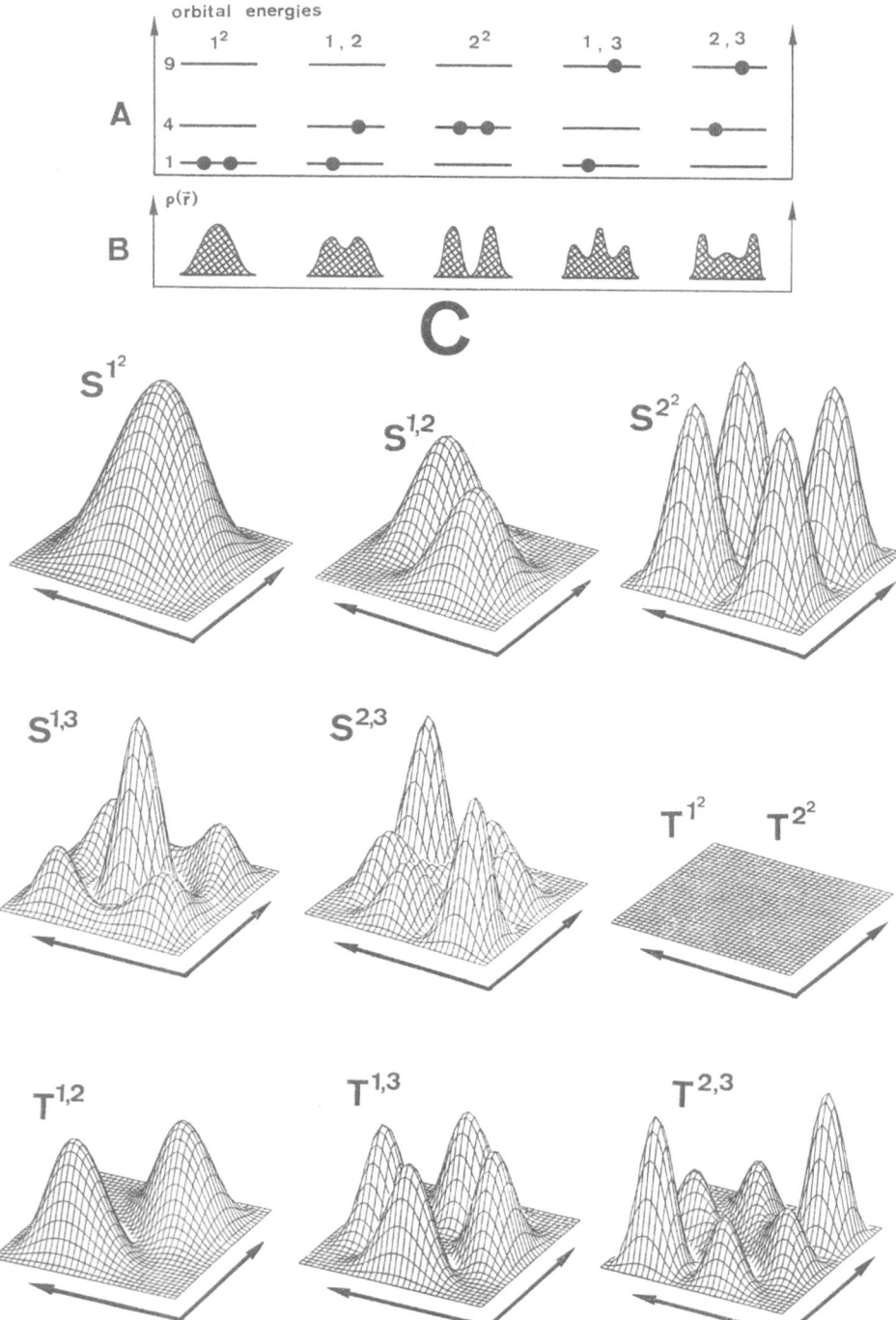

Fig. 6. A) Orbital representation; B) Electron densities $\varrho(\vec{r})$; C) Electron pair densities $\varrho(\vec{r}_1, \vec{r}_2)$ of the singlet and triplet functions of two electrons in a one-dimensional box. The two axes are r_1 and r_2

triplet configuration energy will always be lower than the corresponding singlet energy. Let us stress again that the energy splitting between singlet and triplet configurations:

$$2 K_{ii'} = 2 \langle \phi_i(1)\, \phi_{i'}(2) | e_o^2/r_{12} | \phi_{i'}(1)\, \phi_i(2) \rangle_{1,2} \tag{26}$$

which leads to the lower energy of the triplet function, is a consequence of the different two-electron properties of the space part of the wave functions. It can be quite significant and may reach several 1000 cm^{-1}.

The Configuration Interaction Problem. Returning now to our example, naphthalene: 25 single excitations, each generating 4 determinants, can be formed. Together with S_0 an eigenvalue problem involving a 101×101 matrix results. This problem can be greatly simplified if, instead of using the determinants $\Delta_k^{i\beta \to i'\beta}$, $\Delta_k^{i\beta \to i'\alpha}$, $\Delta_k^{i\alpha \to i'\beta}$, and $\Delta_k^{i\alpha \to i'\alpha}$, we employ the functions $^{3,1}\Phi_k$, $^{3,0}\Phi_k$, $^{3,-1}\Phi_k$, $^{1,0}\Phi_k$ as trial functions (Fig. 7).

Since they are correct eigenfunctions of S^2 and S_z, all matrix elements involving different spin quantum numbers will vanish:

$$\langle ^{S,M_S}\Phi_k | H | ^{S',M'_S}\Phi_e \rangle = \delta_{S,S'}\, \delta_{M_S,M'_S} \langle ^{S,M_S}\Phi_k | H | ^{S,M_S}\Phi_e \rangle \tag{27}$$

The configuration interaction problem decomposes into a 25×25 singlet problem and three completely identical 25×25 triplet problems. In all future calculations it will therefore be sufficient to consider separately the singlet problem and a triplet problem that corresponds to a specific value of M_S.

$$^{S,M_s}\Psi_n(1 \ldots n) = \sum_k {}^{S,M_s}\Phi_k(1 \ldots n)\, a_{kn} \tag{28}$$

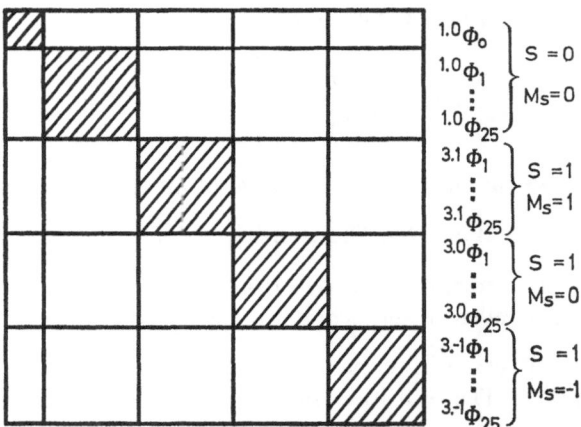

Fig. 7. Configuration interaction problem of dimension 101×101 involving all 25 single excitations in naphthalene

Further simplification in both the singlet and the triplet problem is introduced by considering the symmetries of the space parts. Naphthalene belongs to the

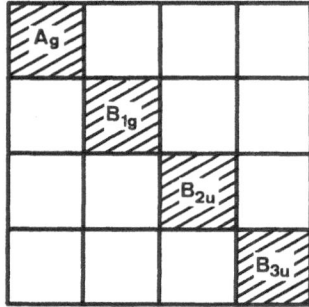

Fig. 8. Block form of the triplet configuration interaction problem involving 25 singly excited configurations (A_g:6; B_{1g}:6; B_{2u}:8; B_{3u}:5). The symmetry classification refers to the space part of the triplet functions

point group D_{2h} and the configuration interaction problem can be grouped in noninteracting blocks of A_g, B_{1g}, B_{2u}, and B_{3u} symmetry (Fig. 8).

$$^{S,M_S}\Psi_n^\Gamma(1\ldots n) = \sum_k {}^{S,M_S}\Phi_k^\Gamma(1\ldots n)\, a_{kn} \tag{29}$$

The excited states having energies below 50,000 cm^{-1} have been obtained from a calculation involving the 25 singly excited configurations and are shown in Fig. 9.

3. Singlet-Singlet Transitions $S_n \leftarrow S_0$

The oscillator strength of an electronic transition from the ground state to an excited singlet state S_n is obtained from [8]

$$f_{n,0} = \frac{8\,\pi^2\, m\, \nu}{3\,h\, e_0^2}\, |\langle\Psi_0|\sum_{i=1}^{n} e_0\, \vec{r}_i|\Psi_n\rangle|^2 \tag{30}$$

where m and e_0 are the mass and the charge of an electron, and $\nu_{n,0}$ is the frequency of the transition $S_n \leftarrow S_0$. The transition dipole moment $\langle\Psi_0|\sum_{i=1}^{n} e_0\, \vec{r}_i|\Psi_n\rangle$ is a measure of the strength of an electronic transition. If the state function $\Psi_n = \sum_t \Phi_t\, a_{tn}$ is expanded in terms of its component configurations, the intensity of a band can be discussed in terms of contributions from the individual participating configurations. The transition dipole moment operator belongs to the class of "one-electron operators" and nonvanishing matrix elements $\langle\Phi_s|\sum_{i=1}^{n} e_0\, \vec{r}_i|\Phi_t\rangle$ will be obtained only if Φ_s and Φ_t differ in not more than one spin-orbital. All singly excited configurations (not considering symmetry restrictions) can therefore contribute to the intensity of $S_n \leftarrow S_0$ transitions and should be included in the configuration interaction problem.

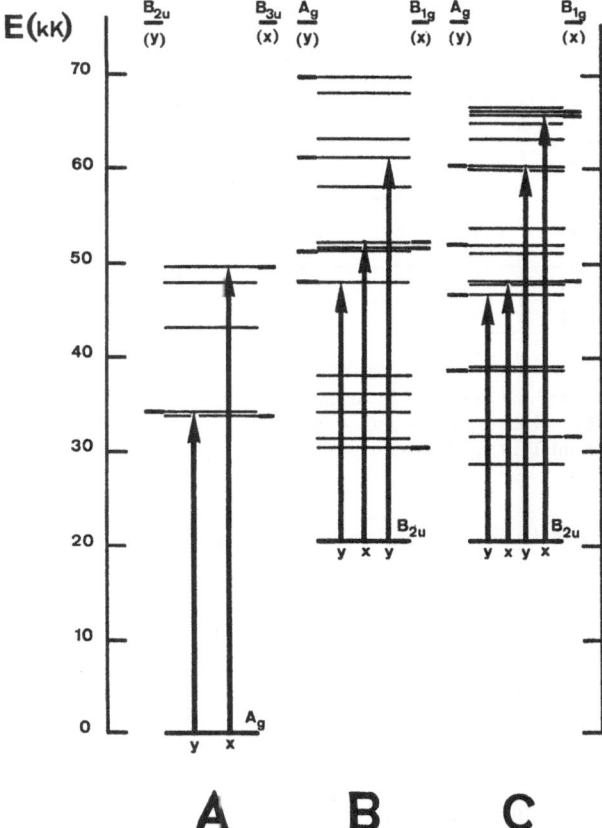

Fig. 9. Calculated $S_n \leftarrow S_0$ and $T_m \leftarrow T_1$ spectra. A) Singlet problem; configuration interaction with 25 singly excited configurations. B) Triplet problem; configuration interaction with 25 singly excited configurations. C) Triplet problem; configuration interaction with 53 configurations

4. Triplet-Triplet Transitions $T_m \leftarrow T_1$

Electronic transitions between the same components of triplet states are spin allowed and have the same order of magnitude as $S_n \leftarrow S_0$ transitions. The transition dipole moment is calculated in the same fashion as before:

$$\langle {}^{3,Ms}\Psi_1 | \sum_{i=1}^{n} e_0 \, \vec{r}_i | {}^{3,Ms}\Psi_m \rangle = \sum_{s} \sum_{t} a_{s1} \, a_{tn} \langle {}^{3,Ms}\Phi_s | \sum_{i=1}^{n} e_0 \, \vec{r}_i | {}^{3,Ms}\Phi_t \rangle \qquad (31)$$

Again, the dipole moment operator connects configurations that differ by not more than one spin orbital. The situation in this case is, however, more complicated. The main configurational component of ${}^{3,Ms}\Psi_1$ is already a singly excited configuration with respect to S_0. In order not to lose any significant contribution to the oscillator strength of the $T_m \leftarrow T_1$ transition, all configurations that are singly excited with respect to the main component of ${}^{3,Ms}\Psi_1$, lying within a reasonable energy gap, should be included in the configuration interaction treatment. Many of these configurations will, of course, be doubly

15

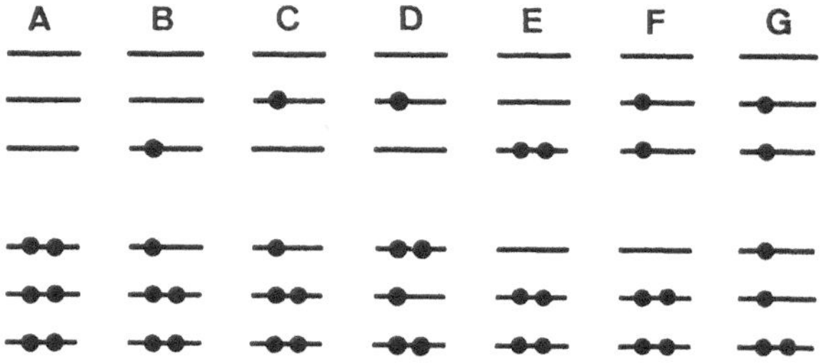

	A	B	C	D	E	F	G
a)	0	1	1	1	2	2	2
b)	1	0	1	2	1	1	1
c)	0	1	1	1	0	1	3

Fig. 10. Types of configurations to be considered in the calculation of $T_m \leftarrow T_1$ spectra. a) Excitation number with respect to the ground state A. b) Excitation number with respect to the main component B of the lowest triplet state T_1. c) Number of triplet functions with a specific M_S value

excited with respect to S_0 and will involve two or four open shells. A schematic representation of the configurations to be considered is given in Fig. 10.

The configurations of type C and D have been discussed before. Type E yields only singlet configurations. Each excitation of type F adds an additional triplet configuration to the CI problem. A total of 16 determinants can be built by filling the four open-shell orbitals in the case G. By linear combinations 2 singlet, 9 triplet, and 5 quintet configurations can be formed. Such a promotion will thus add 3 energetically different triplet configurations having a specific M_S value to the CI problem. The result [9] of such an extended treatment containing a total of 53 configurations is also included in Fig. 9.

5. Singlet-Triplet Transitions $T_m \leftarrow S_0$

The evaluation of the transition dipole moment between the triplet and singlet wave functions considered so far

$$\langle ^{3,M_S}\Psi_m(1\ldots n)\,|\sum_{i=1}^{n} e_0\,\vec{r}_i\,|\, ^{1,0}\Psi_n(1\ldots n)\rangle_{1\ldots n} = 0 \qquad (32)$$

always gives zero, because of the orthogonality of the spin functions. The singlet–triplet and triplet–singlet transitions are spin-forbidden and not accounted for in the approximation of a spin-free Hamiltonian used up to now. If relativistic effects are considered and the "Pauli" approximation of the Hamilton operator is taken, five additional terms have to be included [10]. For our present purpose, just two of them — the interaction between two electronic spins and the interaction between the electron spin and its orbital angular momentum — are important. One may rightfully ask if it is reasonable to include such small terms in a treatment where rather crude approximations have been introduced before. Indeed, the energy corrections introduced by these small perturbation terms are completely insignificant on an absolute basis. The small energy splitting between the triplet sublevels and the small coupling between the singlet and triplet manifold can, however, be determined rather accurately even with approximate wavefunctions. These small terms are thus essential for the description of the zero-field splitting of the triplet states and the oscillator strengths of singlet–triplet transitions.

A. Spin–Spin Interaction

For simplicity, let us return to the true isoenergetic triplet functions given by Eq. (23) involving just the two electrons outside the closed-shell part.

$$^{3,1}\Phi = \frac{1}{\sqrt{2}} \left(\phi_i(1)\, \phi_{i'}(2) - \phi_{i'}(1)\, \phi_i(2) \right) \alpha(1)\, \alpha(2) \qquad\qquad = \bar{\phi}(1,2)\, \vartheta^1(1,2)$$

$$^{3,0}\Phi = \frac{1}{\sqrt{2}} \left(\phi_i(1)\, \phi_{i'}(2) - \phi_{i'}(1)\, \phi_i(2) \right) \frac{1}{\sqrt{2}} \left(\alpha(1)\, \beta(2) + \beta(1)\, \alpha(2) \right) = \bar{\phi}(1,2)\, \vartheta^0(1,2)$$

$$^{3,-1}\Phi = \frac{1}{\sqrt{2}} \left(\phi_i(1)\, \phi_{i'}(2) - \phi_{i'}(1)\, \phi_i(2) \right) \beta(1)\, \beta(2) \qquad\qquad = \bar{\phi}(1,2)\, \vartheta^{-1}(1,2)$$

$$(33)$$

The perturbation $\boldsymbol{H}_{SS}(1,2)$, which describes the interaction of the magnetic dipole moments associated with the spins of two electrons, is given by:

$$\boldsymbol{H}_{SS}(1,2) = g^2\, \beta^2/\hbar^2 \left\{ \frac{\boldsymbol{s}(1)\,\boldsymbol{s}(2)}{r_{12}^3} - 3\, \frac{(\vec{r}_{12} \cdot \boldsymbol{s}(1))\,(\vec{r}_{12} \cdot \boldsymbol{s}(2))}{r_{12}^5} \right\}$$

$$= g^2\, \beta^2/\hbar^2\, r_{12}^5 \left\{ \begin{pmatrix} \boldsymbol{s}_x(1) & \boldsymbol{s}_y(1) & \boldsymbol{s}_z(1) \end{pmatrix} \begin{pmatrix} r_{12}^2 - 3x_{12}^2 & -3x_{12}\,y_{12} & -3x_{12}\,z_{12} \\ -3y_{12}\,x_{12} & r_{12}^2 - 3y_{12}^2 & -3y_{12}\,z_{12} \\ -3z_{12}\,x_{12} & -3z_{12}\,y_{12} & r_{12}^2 - 3z_{12}^2 \end{pmatrix} \begin{pmatrix} \boldsymbol{s}_x(2) \\ \boldsymbol{s}_y(2) \\ \boldsymbol{s}_z(2) \end{pmatrix} \right\}$$

$$(34)$$

where $\beta = e_0\hbar/2mc$ is the Bohr magneton and g the g factor of an electron. If we consider the operator $\boldsymbol{H}_{SS}(1,2)$ as a perturbation to $\boldsymbol{H}(1,2)$, stationary perturbation theory requires the calculation of the matrix elements of the perturbation operator $\boldsymbol{H}_{SS}(1,2)$ in terms of the zero-order triplet functions:

$$\langle \bar{\phi}(1,2)\, \vartheta^i(1,2) \,|\, \boldsymbol{H}_{SS}(1,2) \,|\, \bar{\phi}(1,2)\, \vartheta^{i'}(1,2) \rangle_{1,2} =$$

$$\langle \vartheta^i(1,2) \, \begin{pmatrix} \boldsymbol{s}_x(1) & \boldsymbol{s}_y(1) & \boldsymbol{s}_z(1) \end{pmatrix} \,|\, H_{SS} \,|\, \begin{pmatrix} \boldsymbol{s}_x(2) \\ \boldsymbol{s}_y(2) \\ \boldsymbol{s}_z(2) \end{pmatrix}\, \vartheta^{i'}(1,2) \rangle_{\text{spin}}$$

with

$$(35)$$

$$H_{SS} = g^2\beta^2 \langle \bar{\phi}(1,2) \left| \frac{1}{r_{12}^5} \begin{pmatrix} r_{12}^2 - 3x_{12}^2 & -3x_{12}\,y_{12} & -3x_{12}\,z_{12} \\ -3y_{12}\,x_{12} & r_{12}^2 - 3y_{12}^2 & -3y_{12}\,z_{12} \\ -3z_{12}\,x_{12} & -3z_{12}\,y_{12} & r_{12}^2 - 3z_{12}^2 \end{pmatrix} \right| \bar{\phi}(1,2) \rangle_{\text{space}}$$

If a molecule has symmetry C_{2v} or higher, the space integrals of form $\langle \bar{\phi}(1,2) \, | x_{12} \, y_{12}/r_{12}^5 | \, \bar{\phi}(1,2) \rangle_{\text{space}}$ vanish by symmetry and the positions of the magnetic axes can be taken along the geometrical axis system. If the geometry is lower, a main axis transformation defines the direction of the magnetic axis system. If one introduces the definitions:

$$D = \frac{3}{4} \, g^2 \, \beta^2 \, \langle \bar{\phi}(1,2) \left| \frac{r_{12}^2 - 3 \, z_{12}^2}{r_{12}^5} \right| \bar{\phi}(1,2) \rangle_{1,2}$$

and

$$E = \frac{3}{4} \, g^2 \, \beta^2 \, \langle \bar{\phi}(1,2) \left| \frac{y_{12}^2 - x_{12}^2}{r_{12}^5} \right| \bar{\phi}(1,2) \rangle_{1,2} \tag{36}$$

one obtains, neglecting the cross terms:

$$H_{\text{SS}} = \begin{pmatrix} D/3 & 0 & E \\ 0 & -2D/3 & 0 \\ E & 0 & D/3 \end{pmatrix} \tag{37}$$

The eigenvectors and eigenvalues of H_{SS} are easily determined and given by:

$$
\begin{aligned}
E_x &= D/3 - E & {}^{3,\Gamma_x}\Psi(1,2) &= \bar{\phi}(1,2) \, 1/\sqrt{2} \, \{\alpha(1) \, \alpha(2) - \beta(1) \, \beta(2)\} = \bar{\phi}(1,2) \, \vartheta^x \\
E_y &= D/3 + E & {}^{3,\Gamma_y}\Psi(1,2) &= \bar{\phi}(1,2) \, i/\sqrt{2} \, \{\alpha(1) \, \alpha(2) + \beta(1) \, \beta(2)\} = \bar{\phi}(1,2) \, \vartheta^y \\
E_z &= -2D/3 & {}^{3,\Gamma_z}\Psi(1,2) &= \bar{\phi}(1,2) \, 1/\sqrt{2} \, \{\alpha(1) \, \beta(2) + \beta(1) \, \alpha(2)\} = \bar{\phi}(1,2) \, \vartheta^z \,.
\end{aligned}
\tag{38}
$$

The spin functions ϑ^x, ϑ^y, and ϑ^z are eigenfunctions of $\boldsymbol{S}_x$, $\boldsymbol{S}_y$, and $\boldsymbol{S}_z$ and have an eigenvalue of 0. The three triplet components belonging to a triplet state are not exactly isoenergetic. The splitting characterized by the parameters D and E is small and lies generally below 1 cm^{-1}. Still, the three sublevels have quite distinct and often widely differing properties and much modern research is directed toward the goal of a detailed characterization of the properties of the individual components. This will become apparent in the discussion of spin–orbit coupling effects.

B. Spin–Orbit Interaction

The main effect of taking spin–orbit interaction into account will be an admixture of singlet character to triplet states and triplet character to singlet states. The spin–orbit coupling Hamiltonian can to a good approximation be described by an effective one-electron operator $\boldsymbol{H}_{\text{SO}}$:

$$\boldsymbol{H}_{\text{SO}} = \frac{e_0^2}{2 \, m^2 \, c^2} \sum_k^N \sum_i^n \frac{Z_{k,\,\text{eff}}}{r_{ik}^3} \, \boldsymbol{l}_i \cdot \boldsymbol{s}_i \,. \tag{39}$$

The molecule is approximated by a set of shielded atoms, each center giving rise to a spherical electric field. $Z_{k,\text{eff}}$ can be determined by the Slater rules, and $\boldsymbol{l}_i$ and $\boldsymbol{s}_i$ are the orbital and spin angular momentum operators.

First-order perturbation theory is then applied to derive the nominal "singlet ground state" and "first excited triplet" functions. Pure spin states are no longer possible.

$$
\text{"}1,0\text{"}\Psi_0 = {}^{1,0}\Psi_0 + \sum_{m} \sum_{\Gamma_s} \frac{\langle {}^{3,\Gamma_s}\Psi_m \,|\, H_{SO} \,|\, {}^{1,0}\Psi_0 \rangle}{{}^{1}E_0 - {}^{3}E_m} \cdot {}^{3,\Gamma_s}\Psi_m
$$

$$
\text{"}3,\Gamma_s\text{"}\Psi_1 = {}^{3,\Gamma_s}\Psi_1 + \sum_{n} \frac{\langle {}^{1,0}\Psi_n \,|\, H_{SO} \,|\, {}^{3,\Gamma_s}\Psi_1 \rangle}{{}^{3}E_1 - {}^{1}E_n} \cdot {}^{1,0}\Psi_n \tag{40}
$$

The transition moment from S_0 to the component Γ_s of T_1 is obtained from:

$$
\langle \text{"}1,0\text{"}\Psi_0 \,|\, \sum_{i=1}^{n} e_0 \vec{r}_i \,|\, \text{"}3,\Gamma_s\text{"}\Psi_1 \rangle =
$$

$$
\sum_{n} \frac{\langle {}^{1,0}\Psi_n \,|\, H_{SO} \,|\, {}^{3,\Gamma_s}\Psi_1 \rangle}{{}^{3}E_1 - {}^{3}E_m} \cdot \langle {}^{1,0}\Psi_0 \,|\, \sum_{i=1}^{n} e_0 \vec{r}_i \,|\, {}^{1,0}\Psi_n \rangle \tag{41}
$$

$$
+ \sum_{m} \frac{\langle {}^{3,\Gamma_s}\Psi_m \,|\, H_{SO} \,|\, {}^{1,0}\Psi_0 \rangle}{{}^{1}E_0 - {}^{3}E_m} \cdot \langle {}^{3,\Gamma_s}\Psi_m \,|\, \sum_{i=1}^{n} e_0 \vec{r}_i \,|\, {}^{3,\Gamma_s}\Psi_1 \rangle
$$

The oscillator strength of a specific ${}^{\Gamma_s}T_1 \leftarrow S_0$ transition and the radiative lifetime of a sublevel ${}^{\Gamma_s}T_1$ can now be calculated from:

$$
f\,({}^{\Gamma_s}T_1 \leftarrow S_0) = \frac{8\pi m \nu}{3 h e_0^2} \,|\langle \text{"}1,0\text{"}\Psi_0 \,|\, \sum_{i=1}^{n} e_0 \vec{r}_i \,|\, \text{"}3,\Gamma_s\text{"}\Psi_1 \rangle|^2
$$

$$
1/\tau_r({}^{\Gamma_s}T_1) = \frac{64\,\pi^4\,\nu^3}{3\,h\,c^3} \,|\langle \text{"}1,0\text{"}\Psi_0 \,|\, \sum_{i=1}^{n} e_0 \vec{r}_i \,|\, \text{"}3,\Gamma_s\text{"}\Psi_1 \rangle|^2 \tag{42}
$$

One should clearly note that the frequency of the $T_1 \leftarrow S_0$ and $T_1 \rightarrow S_0$ transitions is governed by the energy of the triplet state T_1, while the intensity and the polarization of these bands are stolen from the $S_n \leftarrow S_0$ and the $T_m \leftarrow T_1$ spectra. A detailed understanding of both, the $S_n \leftarrow S_0$ and the $T_m \leftarrow T_1$ absorption spectra is thus a prerequisite for the understanding of the $T_1 \leftarrow S_0$ and $T_1 \rightarrow S_0$ spectra:

Two further points should be mentioned:

i) The transitions ${}^{\Gamma_s}T_1 \leftarrow S_0$ to the individual sublevels can generally not be resolved, and the total oscillator strength of $T_1 \leftarrow S_0$ is given by:

$$
f = \sum_{\Gamma_s} f\,({}^{\Gamma_s}T_1 \leftarrow S_0) \tag{43}
$$

ii) The three independent exponential phosphorescence decays with radiative lifetimes $\tau_r({}^{\Gamma_s}T_1)(s = x,y,z)$ can only be observed at very low temperature. Above 10 K the three sublevels can no longer be considered isolated. Spin–

19

lattice relaxation becomes fast with respect to the phosphorescence lifetimes, and only a single exponential is observed with lifetime

$$1/\tau_{\mathbf{r}} = \frac{1}{3} \sum_{\Gamma_s} 1/\tau_{\mathbf{r}}(^{\Gamma_s}T_1) \qquad (44)$$

Ellis, Squire, and Jaffe [11] have carried out a detailed study of the spin–orbit coupling in formaldehyde and azulene. The wave functions were calculated using the CNDO/S method, which is more general than the PPP method discussed earlier and also allows σ-orbitals to be included. Let us review their findings on formaldehyde:

Formaldehyde belongs to the symmetry group C_{2v} and the electronic states can be classified into the four irreducible representations A_1, A_2, B_1, and B_2.

Table 2. Character Table C_{2v}

C_{2v}		E	C_2	σ_v	$\sigma_{v'}$
A_1	z	1	1	1	1
A_2	R_z	1	1	—1	—1
B_1	x, R_y	1	—1	1	—1
B_2	y, R_x	1	—1	—1	1

Table 3. Electronic states of formaldehyde

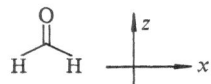

Singlet manifold					Triplet manifold					
							Symmetry			
State	Sym.	eV[1]	f[1]	Pol.	State	Space	Space + Spin	eV[1]	f[1]	Pol.
S_0	A_1	0	—	—						
S_1	A_2	3.5	—	—	T_1	A_2	$B_1/B_2/A_1$	3.5	—	—
S_2	B_1	8.6	0.009	x	T_2	A_1	$B_2/B_1/A_2$	5.4	—	—
S_3	A_1	9.8	0.315	z	T_3	B_1	$A_2/A_1/B_2$	8.6	0.012	y
S_4	B_2	10.0	0.144	y	T_4	B_2	$A_1/A_2/B_1$	8.9	0.090	x
S_5	B_1	11.4	—	—	T_5	B_1	$A_2/A_1/B_2$	11.4	—	—

[1] Ref. [11].

The calculated state energies, the transition moments, and the symmetry classification are given in Table 3. The symmetry species of the triplet functions is obtained by taking the direct product of irreducible representation of the space and the spin functions Γ_x, Γ_y, Γ_z, which transform as the rotations R_x, R_y, and R_z.

Some of the largest spin–orbit coupling matrix elements are given in Table 4.

Table 4. Spin–orbit matrix elements in formaldehyde

$\langle {}^{1,0}\Psi_n \,|\, H_{SO} \,|\, {}^{3,\Gamma_s}\Psi_1 \rangle \cdot i^{-1}$		[1]	$\langle {}^{3,\Gamma_s}\Psi_m \,|\, H_{SO} \,|\, {}^{1,0}\Psi_0 \rangle \cdot i^{-1}$		[1]
Γ_x	$n = 4$	10.3 —	Γ_x	$m = 3$	55.7
	$n = 10$	14.0		$m = 21$	19.1
	$n = 2$	−53.2		$m = 4$	16.8
Γ_y	$n = 12$	−13.3	Γ_y	$m = 10$	32.5
	$n = 21$	16.1		$m = 13$	29.3
				$m = 19$	−33.2
	$n = 0$	−29.5			
Γ_z	$n = 3$	50.7	Γ_z	$m = 1$	29.5
	$n = 7$	− 9.8			

[1] Ref. [11] units cm^{-1}.

The perturbations from several different states as well as from the ground state are of the same order of magnitude and the idea of a "single perturbing state" though often suggested, seems not to be justified. Still, some conclusions based on group-theoretical arguments can be drawn: the spin–orbit operator H_{SO} is totally symmetric and only matrix elements in which both wave functions belong to the same irreducible representation will not vanish (see Table 5).

The phosphorescence from a triplet sublevel will thus be uniquely polarized; the calculated oscillator strengths and transition moments are given in Table 7.

Table 7. Formaldehyde

Triplet level	Polarization	Oscillator strength[1]	Radiative lifetimes[1]
${}^{3,\Gamma_x}\Psi_1(B_1 = A_2 \otimes B_2)$	x	5.1×10^{-11}	36.7 sec
${}^{3,\Gamma_y}\Psi_1(B_2 = A_2 \otimes B_1)$	y	6.8×10^{-8}	0.030 sec
${}^{3,\Gamma_z}\Psi_1(A_1 = A_2 \otimes A_2)$	z	1.7×10^{-7}	0.010 sec

[1] Ref. [11].

The radiative lifetime above 10 K can now be calculated from Eq. (42) and is 0.025 sec, which is a typical value for ${}^3(n, \pi^*)$-carbonyl triplet states.

The space parts of (π,π^*) states are of A_1 or B_1; those of (σ,π^*) or (π,σ^*) states are of A_2 or B_2 symmetry. A closer inspection of the magnitude of the spin–orbit matrix elements, retaining only one-center contributions, shows that $\langle \Psi_{\pi\pi^*} \,|\, H_{SO} \,|\, \Psi_{\pi\pi^*} \rangle = 0$.

Table 5. Formaldehyde

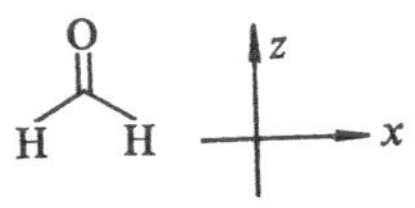

The state	couples with	and steals intensity from	with polarization
$^{3,\Gamma}x\Psi_1(B_1 = A_2 \otimes B_2)$	$^{1,0}\Psi_n(B_1)$	$^{1,0}\Psi_n(B_1) \leftarrow {}^{1,0}\Psi_0(A_1)$	x
$^{3,\Gamma}y\Psi_1(B_2 = A_2 \otimes B_1)$	$^{1,0}\Psi_n(B_2)$	$^{1,0}\Psi_n(B_2) \leftarrow {}^{1,0}\Psi_0(A_1)$	y
$^{3,\Gamma}z\Psi_1(A_1 = A_2 \otimes A_2)$	$^{1,0}\Psi_n(A_1)$	$^{1,0}\Psi_n(A_1) \leftarrow {}^{1,0}\Psi_0(A_1)$	z
$^{1,0}\Psi_0(A_1)$	$^{3,\Gamma}x\Psi_m(A_1 = B_1 \otimes B_1)$	$^{3,\Gamma}x\Psi_m(A_1 = B_1 \otimes B_1) \leftarrow {}^{3,\Gamma}x\Psi_1(B_1 = A_2 \otimes B_2)$	x
$^{1,0}\Psi_0(A_1)$	$^{3,\Gamma}y\Psi_m(A_1 = B_2 \otimes B_2)$	$^{3,\Gamma}y\Psi_m(A_1 = B_2 \otimes B_2) \leftarrow {}^{3,\Gamma}y\Psi_1(B_2 = A_2 \otimes B_1)$	y
$^{1,0}\Psi_0(A_1)$	$^{3,\Gamma}z\Psi_m(A_1 = A_2 \otimes A_2)$	$^{3,\Gamma}z\Psi_m(A_1 = A_2 \otimes A_2) \leftarrow {}^{3,\Gamma}z\Psi_1(A_1 = A_2 \otimes A_2)$	z

Table 6. Azulene

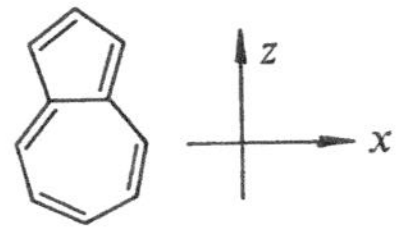

The state	couples with	and steals intensity from	with polarization
$^{3,\Gamma}x\Psi_1(A_2 = B_1 \otimes B_2)$	$^{1,0}\Psi_n(A_2)$	$^{1,0}\Psi_n(A_2) \leftarrow {}^{1,0}\Psi_0(A_1)$	forbidden
$^{3,\Gamma}y\Psi_1(A_1 = B_1 \otimes B_1)$	$^{1,0}\Psi_n(A_1)$	$^{1,0}\Psi_n(A_1) \leftarrow {}^{1,0}\Psi_0(A_1)$	z
$^{3,\Gamma}z\Psi_1(B_2 = B_1 \otimes A_2)$	$^{1,0}\Psi_n(B_2)$	$^{1,0}\Psi_n(B_2) \leftarrow {}^{1,0}\Psi_0(A_1)$	y
$^{1,0}\Psi_0(A_1)$	$^{3,\Gamma}x\Psi_m(A_1 = B_2 \otimes B_2)$	$^{3,\Gamma}x\Psi_m(A_1 = B_2 \otimes B_2) \leftarrow {}^{3,\Gamma}x\Psi_m(A_2 = B_1 \otimes B_2)$	forbidden
$^{1,0}\Psi_0(A_1)$	$^{3,\Gamma}y\Psi_m(A_1 = B_1 \otimes B_1)$	$^{3,\Gamma}y\Psi_m(A_1 = B_1 \otimes B_1) \leftarrow {}^{3,\Gamma}y\Psi_m(A_1 = B_1 \otimes B_1)$	z
$^{1,0}\Psi_0(A_1)$	$^{3,\Gamma}z\Psi_m(A_1 = A_2 \otimes A_2)$	$^{3,\Gamma}z\Psi_m(A_1 = A_2 \otimes A_2) \leftarrow {}^{3,\Gamma}z\Psi_m(B_2 = B_1 \otimes A_2)$	y

Quite generally, El-Sayed [13,14] derived the following rules for spin–orbit coupling:

$$(n, \pi^*) \longleftrightarrow (\pi, \pi^*) \qquad (\pi, \pi^*) \nleftrightarrow (\pi, \pi^*) \qquad (n, \pi^*) \nleftrightarrow (n, \pi^*) \qquad (45)$$

In azulene the coupling of $^{3,\Gamma_y}\Psi_1^{\pi,\pi^*}(A_1 = B_1 \otimes B_1)$ to $^{1,0}\Psi_n^{\pi,\pi^*}(A_1)$ and of $^{1,0}\Psi_0^{\pi,\pi^*}(A_1)$ to $^{3,\Gamma_y}\Psi_m^{\pi,\pi^*}(A_1 = B_1 \otimes B_1)$ can thus be neglected and the only remaining contribution leads exclusively to y or out-of-plane polarization, a result which can be generalized to all planar aromatic hydrocarbons.

C. Vibronic Interaction

Non-totally symmetric vibrations lower the symmetry of a molecule and previously forbidden bands may become allowed. The Hamiltonians considered up to now were all given for a fixed nuclear equilibrium geometry. A Taylor series expansion in the normal coordinates Q around this nuclear equilibrium geometry gives

$$H(Q) = H^0 + H_{SO}^0 + \sum_k \left(\frac{\partial H}{\partial Q_k}\right)_0 Q_k + \sum_k \left(\frac{\partial H_{SO}}{\partial Q_k}\right)_0 Q_k \qquad (46)$$

Singlet–triplet mixing by first-order perturbation theory can now be introduced by two mechanisms:

 I: direct spin–orbit coupling H_{SO}

 II: spin–vibronic coupling $\sum_k \left(\frac{\partial H_{SO}}{\partial Q_k}\right) Q_k$

Second-order perturbation theory gives rise to two additional mechanisms involving an intermediate state that is vibrationally coupled to one and spin–orbit coupled to the other manifold (Fig. 11).

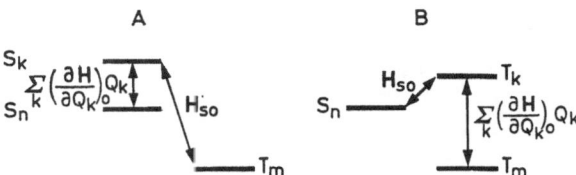

Fig. 11 A. Spin-orbit coupling with vibronic interaction in the singlet manifold. Mechanism III
Fig. 11 B. Spin-orbit coupling with vibronic interaction in the triplet manifold. Mechanism IV

The mechanisms II, III, and IV introduced by vibronic interactions can become significant whenever the direct spin–orbit route is weak or symmetry-forbidden.

Let us conclude this section by referring to some additional literature, which can be recommended for the reader interested in this field [15–25].

III. Triplet-Triplet Absorption Spectra

A review of all the different methods used to investigate triplet absorption spectra and a compilation of all reliable data up to 1972 has recently been given by Labhart and Heinzelmann [26]. Here we discuss the three main experimental methods used in triplet-triplet spectroscopy and review some of the important papers.

1. Flash Photolysis

A typical arrangement for flash photolysis experiments is shown in Fig. 12. A high-voltage capacitor that can store several 100 J is discharged through a flash tube, giving an intensive light pulse of several μsec duration. The high concentration of intermediates formed is studied either by flash or kinetic spectroscopy.

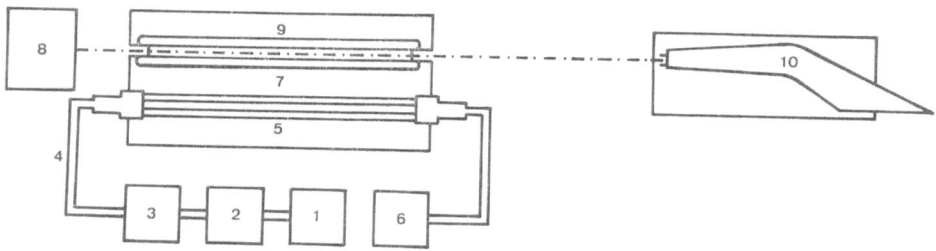

Fig. 12. A flash photolysis apparatus. 1, high-voltage power supply; 2, 10 MΩ resistor; 3, high-voltage capacitor; 4, coaxial cable; 5, flash tube; 6, vacuum system; 7, reflector; 8, pulsed spectroscopic light source; 9, measuring cell; 10, Hilger medium quartz spectrograph. (From Vallotton and Wild, Ref.[27])

In flash spectroscopy a second spectroscopic flash is fired a short time after the photolysis flash and the transient absorption spectrum is registered on a photographic plate (Fig. 13). Repeating the experiments with different delay times gives complete information about the wavelength and the time behavior of the intermediate absorptions.

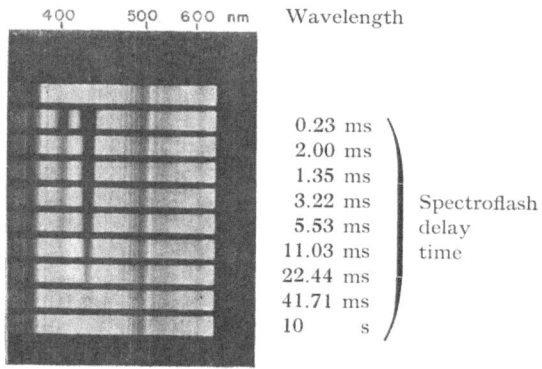

Fig. 13. Triplet-triplet absorption spectrum of anthracene recorded in plexiglas at room temperature (From Wild, Ref.[28])

In kinetic spectroscopy a continuous light source is used. The transient signal at a fixed wavelength is detected with a photomultiplier and displayed on a storage oscilloscope. Repeating the kinetic experiments at several wavelength positions again allows us to determine transient absorption spectra.

The pioneering work on triplet–triplet absorption studies by Porter and Windsor [29] has been followed up by many authors. Some of the best spectra of polyacene were recently published by Meyer, Astier, and Leclercq [30]. Their samples were contained in a low-temperature quartz cell (length 1—210 mm) of 1 cm diameter. Movable Suprasil pistons acted as cell windows and followed the concentration of the sample during cooling, thus eliminating the need for a low temperature concentration correction. Almost complete conversion to the triplet state was achieved for anthracene and tetracene. The $S_n \leftarrow S_0$ and $T_m \leftarrow T_1$ spectra of anthracene observed at 113 K in alcohol are shown in Fig. 14. They were calculated from the densitometer curves of the photographic plates. All triplet bands were verified by measuring their decay times, which had to agree. Such a check allowed the elimination of signals resulting from other species, such as radicals produced by photoionization.

Some of the authors' conclusions are:

i) The width of the vibronic bands depends mostly on the energy above the ground state. The nature of the transitions $S_n \leftarrow S_0$ or $T_m \leftarrow T_1$ is not important.

ii) More electronic transitions can be seen in the $T_m \leftarrow T_1$ than in the $S_n \leftarrow S_0$ spectrum.

iii) The absorption spectrum of T_1 is not strongly influenced by methyl substitution. It seems therefore justified to apply standard π-electron methods to the study of excited-state absorption. The assignment of the individual levels will be discussed later.

2. Modulation Excitation Spectroscopy

The high sensitivity of lock-in amplification can be applied to detect the small periodic changes in transmittance caused by modulated excitation with UV light. Measurements of the amplitude and phase shift of the response signal allow us to determine the spectra and lifetimes of the transient species.

One of the first applications of this chopped-beam irradiation technique to the measurement of triplet spectra was reported by Labhart [31]. From a knowledge of the intensity of the irradiation light, he determined the quantum yield of triplet generation to be 0.55 ± 0.11 for outgassed solutions of 1,2-benzanthrazene in hexane at room temperature. Hunziker [32] has applied this method to the study of the gas-phase absorption spectrum of triplet naphthalene. A gas mixture of 500 torr N_2, 0.3 mtorr Hg, and about 10 mtorr naphthalene was irradiated by a modulated low-pressure mercury lamp. The mercury vapor in the cell efficiently absorbed the line spectrum of the lamp and acted as a photosensitizer. The triplet state of naphthalene was formed directly through collisional deactivation of the excited mercury atoms.

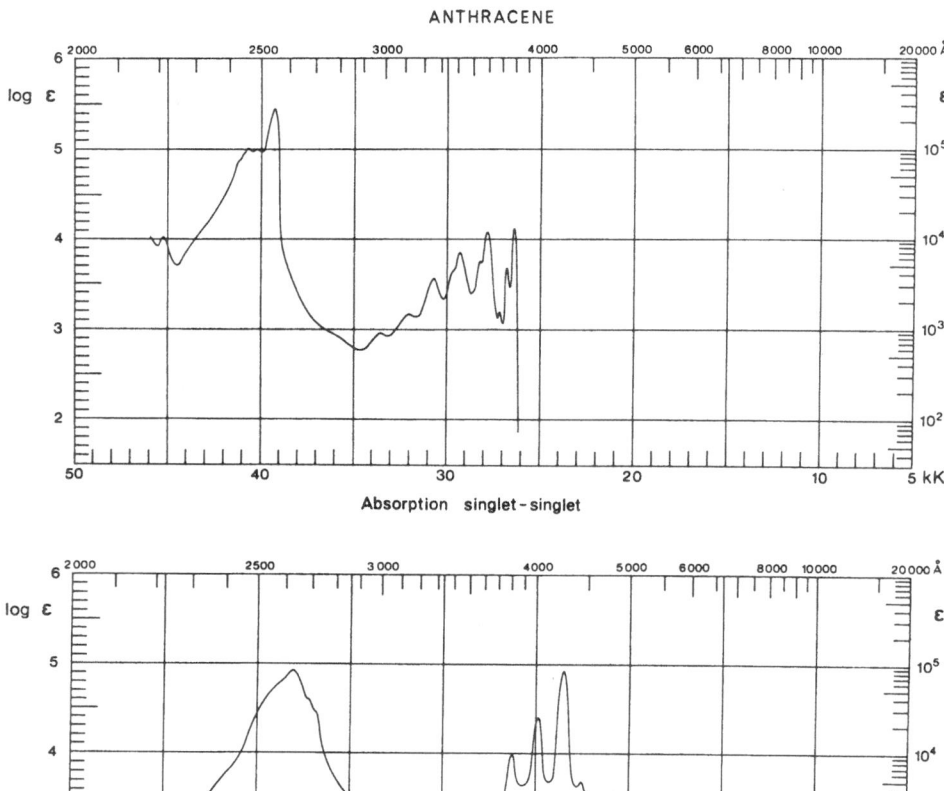

Fig. 14. $S_n \leftarrow S_0$ and $T_m \leftarrow T_1$ absorption spectra of anthracene in alcohol at 113 K (From Meyer, Astier, and Leclercq, Ref.[30])

3. The Stationary-State Method

The rather long lifetimes of many triplet states often allow an appreciable build-up of the triplet-state population under constant illumination. Already in 1954 Craig and Ross [33] measured reliable $T-T$ spectra on a single-beam recording spectrophotometer. The observation of new absorption bands in the visible part of the spectrum is not difficult and there is satisfactory agreement between the peak positions reported from different laboratories. The determination of triplet extinction coefficients requires additionally a knowledge of the stationary triplet concentration, which is much harder to obtain. The corresponding literature values are widely scattered.

Simultaneous measurements of the $T-T$ spectrum and of the $(\Delta m = 2)$ ESR signal have been performed by Brinen [34] under steady-state excitation. The triplet concentration can be determined independently from the spectral measurements by comparing the integrated $(\Delta m = 2)$ signal with a calibration radical signal.

A spectrophotometer which allows spectroscopic and kinetic measurements to be made on a light irradiated sample has been developed by Ranalder et al. [35]. The instrument is completely controlled by a small PDP-8/I computer. Great flexibility is introduced through software control. Several data collection routines have been written, and methods for determining molar absorption coefficients of metastable states have been discussed.

The method of photoselection to study the polarization of triplet–triplet transitions has been applied by El-Sayed and Pavlopoulos [36] to several polyacenes. Let us discuss some of the results obtained on napthalene, where the x-axis has been chosen along the long molecular axis and the y-axis along the short one. The first very weak $S_1 \leftarrow S_0$ absorption band ($^1B_{3u}^- \leftarrow {}^1A_g^-$, x-polarized) is vibrationally coupled to the nearby $^1B_{2u}^+$ S_2 state and gains y polarization. Irradiation into the band system S_1 and S_2 is therefore expected to introduce rather low but definite y-axis polarization. All three strong $T_m \leftarrow T_1$ bands around 4000 Å showed equal and negative polarization degrees and were thus assigned to the x-polarized transition of type $^3B_{1g}^- \leftarrow {}^3B_{2u}^+$. The spacing of about 1500 cm^{-1} between the bands corresponds to a known skeletal vibration. Calculations indeed predict an x-polarized $T_m \leftarrow T_1$ transition around 4000 Å. In addition, a second, somewhat weaker transition of y polarization is expected in this range. Better wavelength resolution of the $T_m \leftarrow T_1$ spectrum revealed five

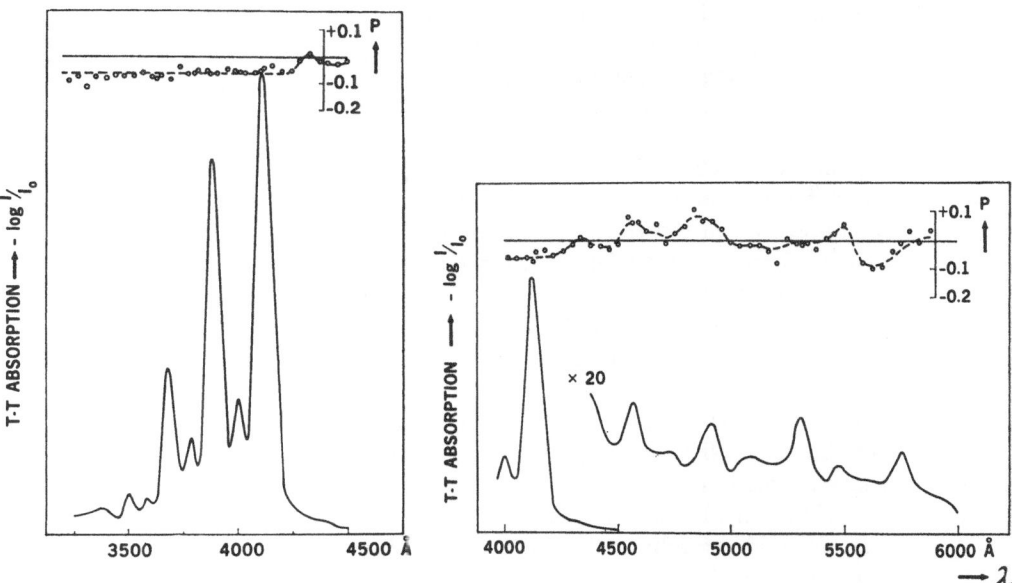

Fig. 15. $T_m \leftarrow T_1$ absorption and polarization spectra of naphthalene-d$_8$ in 3-methylpentane at 77 K (From Pavlopoulos, Ref.[37])

new weak bands and motivated Pavlopoulos [37] to reinvestigate the polarization properties. Again, all bands showed negative values of polarization, and it is now accepted that they all belong to the same electronic transition.

The investigation of the 4300—6000 Å region revealed four weak bands at 4560, 4916, 5312, and 5750 Å which have positive and negative polarization values. It is now believed that they all belong to the expected $^3A_g^- \leftarrow {}^3B_{2u}^+$ transition, the negative polarization being introduced through vibronic coupling to the nearby $^3B_g^-$ state. It should be mentioned here that this transition does not correspond to the $^3A_g^- \leftarrow {}^3B_{2u}^+$ predicted by Orloff [38], having an oscillator strength of 0.58. He used only first-order configuration interaction, and his reported $^3A_g^-$ state is actually the second state of $^3A_g^-$ symmetry. The first $^3A_g^-$ state is described mostly by the orbital promotions $(4 \to 8)-(3 \to 7)$. It is doubly excited with respect to the lowest triplet state and should not carry oscillator strength. The nonzero oscillator strength is only introduced by configuration interaction and therefore depends strongly on the size of configuration interaction included. Still, most calculations seem to overemphasize the strength of the y-polarized $T_m \leftarrow T_1$ transitions.

The polarization study on naphthalene was complemented by Lavalette [39] who determined the polarized excitation spectrum, again using photoselection. The polarization of the strong $T_m \leftarrow T_1$ band at 4170 Å was monitored as a function of the wavelength of polarized excitation into the singlet bands. As expected, a minimum polarization value of —0.18 was obtained at 2900 Å near the 0—0 band of the $S_2 \leftarrow S_0$ ($^1B_{2u}^+ \leftarrow {}^1A_g^+$) transition.

4. The Calculation of Triplet-Triplet Spectra

The pioneering work of Pariser and Parr [1,2] opened the route to the discussion of $S_n \leftarrow S_0$ and $T_m \leftarrow T_1$ absorption spectra of large aromatic molecules. It is to be noted that, using semiempirical methods, one can only expect to describe the few lowest electronic states to an acceptable degree of agreement with experiment. The calculation of triplet–triplet transitions, starting at the already rather high-lying energy level T_1, will contain more transitions in the visible and the UV range up to 50,000 cm^{-1} than will the corresponding $S_n \leftarrow S_0$ spectra. Even though less precision in the triplet calculations must be expected, partial or full assignment has been achieved for several molecules.

An assignment of the triplet states in cata-condensed hydrocarbons based on the perimeter model by Moffitt [40] was given by Kearns [41]. The free-electron model was used by Nouchi [42] to classify the higher triplet states. De Groot and Hoijtink [43] included all configurations within 7.5 eV that were singly or doubly excited with respect to T_1 in a Pariser-Parr-type configuration interaction problem. They were able to predict correctly the strongest band of the $T_m \leftarrow T_1$ spectrum of naphthalene, which lies at 44,000 cm^{-1}. Their work removed the discrepancies between previous calculations and the experimental facts and it forms the basis of our present understanding of triplet–triplet spectra. The very close correspondence between $T_m \leftarrow T_1$ spectra and the absorption spectrum of dinegative ions was pointed out by Hoijtink [44]. Orloff [38] calculated transition energies and oscillator strengths of 12 alternant hydrocarbons, using just first-order configuration interaction between degenerate states. This seems to be a very crude approximation in view of the heavy configuration mixing involved between higher triplet states. All $T_m \leftarrow T_1$ transitions were overestimated by 0.4 eV.

Finally, the $S_n \leftarrow S_0$ and $T_m \leftarrow T_1$ absorption spectra of 15 conjugated hydrocarbons were calculated by Pancir and Zahradnik [45,46] using a modified semiempirical parameter set. Their description of the singlet–triplet and triplet–triplet spectra achieves about the same accuracy as the PPP method for singlet–singlet spectra.

IV. Singlet-Triplet Absorption Spectra

The very weak $T_m \leftarrow S_0$ transitions are hard to observe directly by absorption spectroscopy. Even with long cells, the high concentrations required present solubility — and what is more important — purity problems. An impurity of $1:10^6$ may give rise to absorption bands which have the same intensity as the expected $T_1 \leftarrow S_0$ absorption. The experimental conditions, therefore, have to be chosen to allow an increase of the $T_1 \leftarrow S_0$ oscillator strength to be achieved through perturbation by paramagnetic species (O_2 or NO) or heavy atoms. Alternatively, an indirect method, phosphorescence excitation spectroscopy, which has high sensitivity and selectivity, may be applied.

A mirror-image relation (Fig. 16) between the $T_1 \leftarrow S_0$ absorption spectrum and the $T_1 \rightarrow S_0$ emission spectrum can be expected whenever there is a relatively small geometry change between ground and first triplet states, and S_0 and T_1 have similar vibrational spectra.

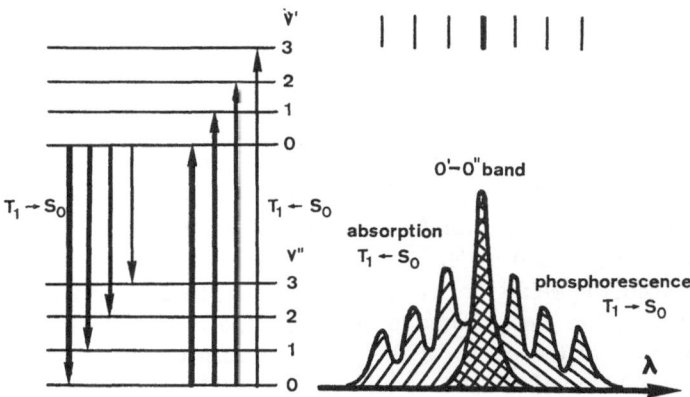

Fig. 16. Mirror-image relation between the $T_1 \leftarrow S_0$ absorption spectrum and the $T_1 \rightarrow S_0$ emission spectrum

1. The Phosphorescense Excitation Method

$T_1 \leftarrow S_0$ spectra can be determined by scanning the sample with monochromatic light and by monitoring the phosphorescence intensity at a fixed wavelength (see also Chapter V). The number of photons absorbed $N_{abs}(\lambda)$ by the sample per unit time is easily obtained as the difference between the rate of photons entering N_{in} and leaving N_{out} the sample:

$$N_{abs}(\lambda) = N_{in}(\lambda) - N_{out}(\lambda) = N_{in}(\lambda)\ (1 - 10^{-\varepsilon(\lambda)\cdot c \cdot l}) \qquad (47)$$

where $\varepsilon(\lambda)$ is the molar extinction coefficient at the excitation wavelength λ, c the concentration, and l the cell length. In the region of $T_1 \leftarrow S_0$ transitions the sample is always optically thin: $\varepsilon(\lambda) \cdot c \cdot l \ll 1$ and the exponential can be expanded:

$$N_{abs}(\lambda) = N_{in}(\lambda) \cdot \ln 10 \cdot \varepsilon(\lambda) \cdot c \cdot l \qquad (48)$$

The total number of phosphorescence photons N_{phos} emitted per unit time is given by:

$$N_{phos}(\lambda) = \eta_{phos}(\lambda)\ N_{abs}(\lambda) \qquad (49)$$

and

$$N_{phos}(\lambda)/N_{in}(\lambda) \propto \varepsilon\ (\lambda)$$

if it is assumed that the quantum yield of triplet formation does not depend on the excitation wavelength. The phosphorescence intensity corrected for the intensity of the monochromatic excitation light is therefore proportional to the molar extinction coefficient of the $T_1 \leftarrow S_0$ transition.

The observation of a phosphorescence signal offers a number of distinct advantages over a direct absorption measurement. Let us first mention the extremely high sensitivity. The method of single photon counting allows one in principle to count the number of the emitted phosphorescence photons. Secondly, high specifity is introduced by the possibility of setting the detecting monochromator at the maximum of the expected phosphorescence, thus eliminating signals from impurities.

The method was first applied by Rothman, Case, and Kearns [47] to the determination of the $T_1 \leftarrow S_0$ absorption spectrum of 1-bromonaphthalene. Sixteen photochemically active aromatic ketones and aldehydes have been investigated by Kearns and Case [48]. Transitions from S_0 to two triplet states were located and assigned as $^3(n,\pi^*)$ and $^3(\pi,\pi^*)$.

Let us consider in detail a recent paper by Jones, Kearns, and Wing [49], which represents one of the most thorough investigations of singlet–triplet transitions and reveals much of the power of the phosphorescence excitation method.

The molecule methyl-5(10)-octalin-1,6-dione (1) was chosen as model compound for the steroidal enones investigated earlier. Oriented single crystals of (1) were studied at low temperature (4 K). First the crystal structure of (1)

(1)

had to be determined, which gave the following important result: the unit cell contains two molecules, which are enantiomers and related by a center of inversion at the center of the unit cell. The directions of the corresponding transition moments are parallel in both molecules; thus no cancellation of polarization information is expected. The four atoms of the enone chromophore, as well as the adjacent atoms, lie almost in a plane.

Two orientations have been chosen (Fig. 17) for polarized excitation spectroscopy. The first is used to distinguish between in-plane and out-of-plane effects, while the second one allows a comparison of the two in-plane directions.

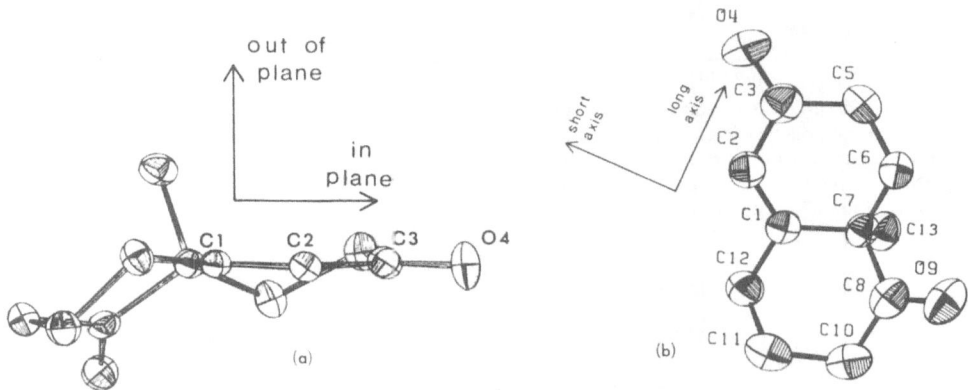

Fig. 17. The structure of the ketone (1) molecule as determined by x-ray crystallography shown in the (approximate) orientations used in the polarization measurements. In (a) the in-plane versus out-of-plane polarization directions are compared. The in-plane extinction direction shown is nearly parallel to the C1—C2 double bond, being tipped just 5° out of plane. In (b) the two directions in plane are compared. The long axis extinction direction shown is exactly in plane and makes an angle of 40° with the C1—C2 double bond. (From Jones, Kearns, and Wing, Ref.[49])

The corresponding polarized excitation spectra are given in Figs. 18 and 19. There is a very weak 0'—0'' peak in the $T_1 \leftarrow S_0$ transition at 407 nm. Toward lower wavelengths the intensity increases while the fine structure disappears and a rather broad maximum is observed at 388 nm.

Let us note that there is a small gap of about 30 cm^{-1} between the 0'—0'' band observed in phosphorescence excitation and phosphorescence emission (Fig. 20). Such an effect is typical for crystals where emission occurs generally from traps that lie 20—50 cm^{-1} below the host level. The phosphorescence spectrum is dominated by the strong 0' → 0'' band, an indication that there is little geometry change between S_0 and T_1.

The lowest triplet state T_1 is of $^3(\pi,\pi^*)$ origin and mixes strongly by direct spin–orbit coupling with the $^1(n,\pi^*)$ state (mechanism I, Chapter II.5.C). Intensity is stolen from the $^1(n,\pi^*) \leftarrow {}^1(\pi,\pi^*)$ transition, which is mostly out-of-plane polarized but also contains a significant in-plane contribution resulting from a

31

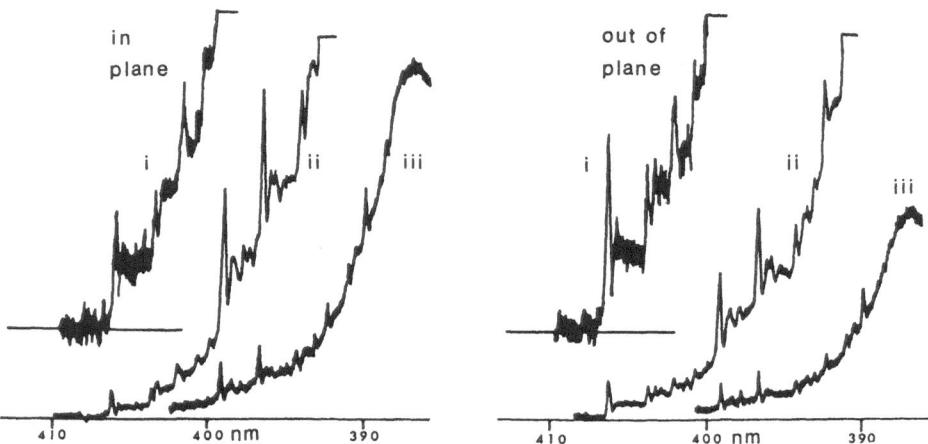

Fig. 18. The polarized excitation spectra in the $T_{\pi\pi^*} \leftarrow S_0$ region comparing the in-plane and out-of-plane polarizations at 4.2 K and slitwidths of 100 μ for curves i, 80 μ for curves ii, and 50 μ for curves iii. For the 0—0 band the out-of-plane polarization is favored 2:1 over the in-plane polarization while the higher energy bands are favored in the in-plane polarization. (From Jones, Kearns, and Wing, Ref.[49])

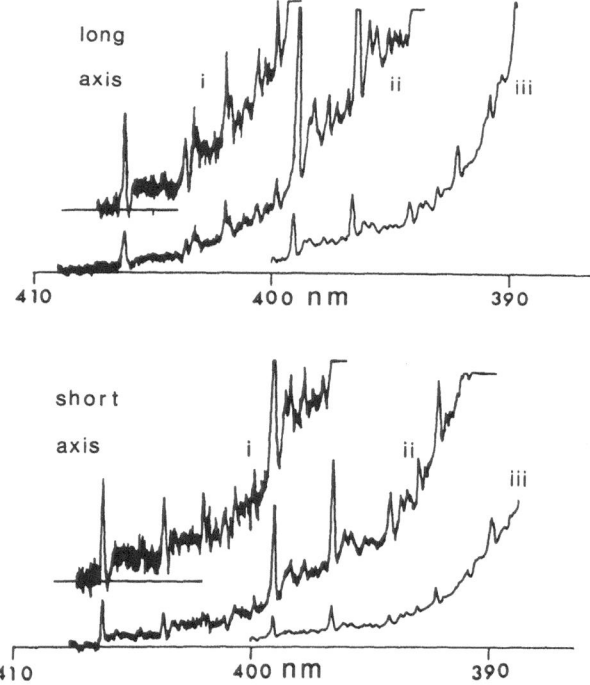

Fig. 19. The polarized excitation spectra in the $T_{\pi\pi^*} \leftarrow S_0$ region comparing the long-axis and short-axis polarizations at 4.2 K and slitwidths for all the curves of 50 μ. The 0—0 band is nearly depolarized while the higher energy bands are favored in the long axis polarization. The gain for curve i is about 5 times that used for curve ii, which is in turn about 5 times that used for curve iii. Comparable curves in the upper and lower spectra are at the same gain. (From Jones, Kearns, and Wing, Ref.[49])

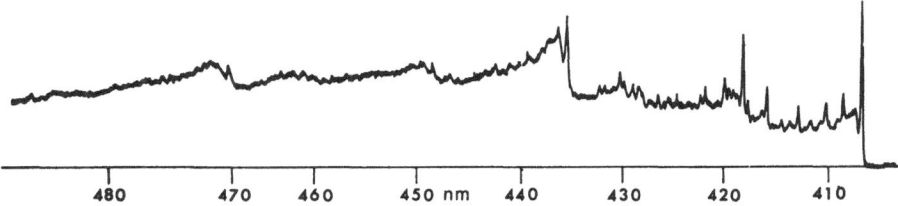

Fig. 20. The phosphorescence emission spectra of a single crystal of ketone (1) at 4.2 K and a slitwidth of 50 μ. (From Jones, Kearns, and Wing, Ref.[49])

distortion of the planar enone chromophore by the out-of-plane hydrogen and methyl groups. The spectra given in Fig. 18 show that out-of-plane polarization dominates by $2:1$ for the first few bands, including the $0' \leftarrow 0''\ T_{\pi,\pi^*} \leftarrow S_0$ transition. The transitions to higher vibrational levels become increasingly in-plane polarized, indicating vibronic mixing with higher $^3(n,\pi^*)$ states.

It is interesting to observe that the mirror-image relation between the $T_1 \leftarrow S_0$ absorption and the $T_1 \rightarrow S_0$ emission is not fulfilled. The two low-lying triplet states $^3(n,\pi^*)$ and $^3(\pi,\pi^*)$ interact strongly by out-of-plane vibrations. The intensity introduced by spin–orbit coupling with vibronic coupling in the triplet manifold (mechanism IV) has the same spin–orbit part but differs in the vibronic coupling elements for the $1' \leftarrow 0''$ band in absorption and the $0' \rightarrow 1''$ band in emission. The ratio of the two intensities is given by:

$$\left\{ |\langle {}^3(n,\pi^*)| \left(\frac{\partial H}{\partial Q_k}\right)_0 Q_k |{}^3(\pi,\pi^*)\rangle / \Delta E_a|^2 \right\} \Big/ \left\{ |\langle {}^3(n,\pi^*)| \left(\frac{\partial H}{\partial Q_k}\right)_0 Q_k |{}^3(\pi,\pi^*)\rangle / \Delta E_b|^2 \right\}$$
$$= (\Delta E_b / \Delta E_a)^2 \tag{50}$$

where Q_k is an out-of-plane normal coordinate. Small spacings between the $^3(\pi,\pi^*)$ and the $^3(n,\pi^*)$ level imply large $(\Delta E_b/\Delta E_a)^2$ ratios, thus the $1' \leftarrow 0''$ band in absorption will be much stronger than the $0' \rightarrow 1''$ in emission (Fig. 21).

Phosphorescence excitation spectroscopy also allows us to observe the transitions starting at 389 nm to the second triplet state, which is of $^3(n,\pi^*)$ nature. Direct spin–orbit coupling (mechanism I) to a $S_n(\pi,\pi^*)$ state introduces strong in-plane, long-axis polarization. Indeed, in-plane polarization is preferred over out-of-plane polarization by $3:1$, and long-axis polarization is about four times stronger than the short-axis contribution.

The intensity of singlet–triplet transitions can be increased by the external and internal heavy-atom effect. It has been noticed by Kearns [48] that the $T_{\pi,\pi^*} \leftarrow S_0$ transitions were enhanced by a factor of about 2 on passing from an ordinary low-temperature glass, such as a $2:1:1$ mixture of ether, ethanol, and toluene, to a heavy-atom glass, such as a $2:2:1:1$ mixture of ethyl iodide, ether, ethanol, and toluene.

The intensity of $T_{n\pi^*} \leftarrow S_0$ transitions remains unaffected by external heavy-atom perturbations. This difference in behavior forms an additional tool for

$$|\langle^3(n, \pi^*)|(\partial H/\partial Q)Q\,|^3(\pi, \pi^*)\rangle/\Delta E_b\,|^2,$$

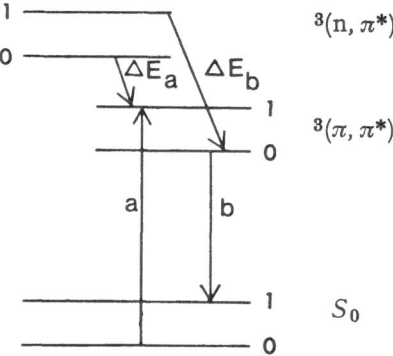

Fig. 21. The energy-level diagram for vibronic coupling. The $1' \leftarrow 0''$ absorption band of the $^3(\pi, \pi^*)$ state gains intensity by route a, and the gain in intensity is proportional to $(\Delta E_a)^{-2}$. The $0' \rightarrow 1''$ emission band gains intensity by route b, and the gain in intensity is proportional to $(\Delta E_b)^{-2}$. Since ΔE_b bis much larger then ΔE_a, the absorption gains more intensity than the emission does. (From Jones, Kearns, and Wing, Ref.[49])

distinguishing experimentally between $^3(n, \pi^*)$ and $^3(\pi, \pi^*)$ states. The internal heavy-atom effect depends very strongly on the atomic number of the "heavy-atom" substituent. The phosphorescence excitation spectra of 1- and 2-chloro-, bromo-, and iodo-naphthalenes and of about 40 other aromatic hydrocarbons have been reported by Marchetti and Kearns [50]. The increase in $T_1 \leftarrow S_0$ intensity is best illustrated by the strong decrease in the phosphorescence lifetimes: 1.5 sec for 1-fluoro-naphthalene, 0.025 sec for 1-iodo-naphthalene.

Hirota [51] used doped crystals to observe weak $T_1 \leftarrow S_0$ absorption spectra by phosphorescence excitation spectroscopy. Triplet excitons of the host are formed by direct light absorption. The guest molecules, chosen to have lower triplet energy, act as traps and emit guest phosphorescence.

In pure crystals, singlet excitons can be created by mutual annihilation of triplet excitons. The intensity of the singlet exciton fluorescence depends quadratically on the triplet exciton concentration and is therefore proportional to the square of the singlet–triplet extinction coefficient. It is interesting to compare such a "delayed fluorescence" excitation spectrum, observed by Avakian *et al.* [52] on naphthalene, with a corresponding phosphorescence excitation spectrum (Fig. 22).

2. Perturbation by Paramagnetic Species, the Oxygen Pressure Method

A very simple and effective method of increasing the intensity of $T_1 \leftarrow S_0$ transitions by several orders of magnitude was introduced by Evans [53–55]. Either the pure liquid or concentrated solutions of an aromatic hydrocarbon in chloroform were saturated with oxygen or nitric oxid at high pressures. The newly appearing absorption bands are proportional to the applied gas pressures from 0—100 atm. The $T_m \leftarrow S_0$ absorptions are in general well structured and the posi-

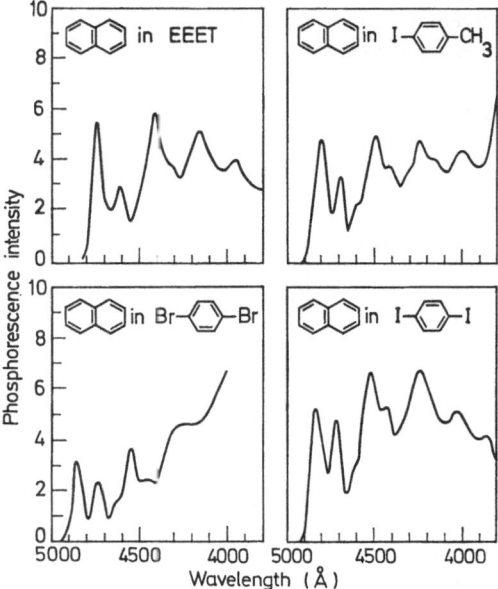

Fig. 22 A. The singlet-triplet absorption spectra of naphthalene in ethyliodide, ether, ethanol, toluene (EEET 2:2:1:1) and in several halogenated benzenes at 77 K. (From Marchetti und Kearns, Ref.[50])

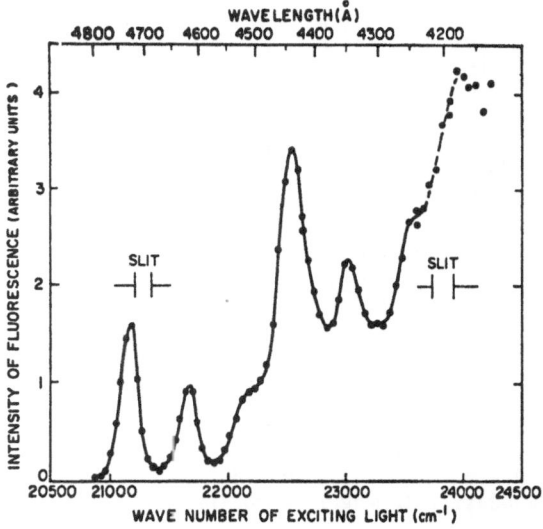

Fig. 22 B. Excitation spectrum at room temperature showing the intensity of delayed fluorescence of a naphthalene crystal as a function of the wavelength of the exciting light. The ordinate is proportional to the square of the singlet-triplet absorption coefficient. (From Avakian and Abramson, Ref.[52])

tions of the long wavelength bands relative to the 0—0 phosphorescence bands given in Table 8 establish clearly the nature of the induced $T_m \leftarrow S_0$ bands. The onset of strong $S_1 \leftarrow S_0$ absorption limits the observation of higher triplet states. In addition, strong unstructured bands due to charge-transfer complexes can cover the $T_1 \leftarrow S_0$ spectrum.

Table 8. Triplet levels (cm^{-1}) of aromatic molecules obtained from oxygen perturbation and phosphorescence spectra, respectively (From Ref. [53])

Compound	Oxygen perturbation (CHCl$_3$ soln.)	Phorsphorescence (low-temp. glass)
Benzene	29 440	29 740
Fluorene	23 580	23 750
Phenanthrene	21 600	21 600
Naphthalene	21 180	21 246
Anthracene	14 870	14 927
1-Bromonaphthalene	20 650	20 700

The oxygen-perturbed singlet–triplet spectra of aromatic carbonyl compounds were investigated by Warwick and Wells [56] (Fig. 23). Transitions to $^3(\pi,\pi^*)$ states were enhanced by the perturbing agent while transitions to $^3(n,\pi^*)$ states remained unaffected. It should be mentioned, however, that Evans [54] also observed an oxygen-perturbed increase of the intensity of the $T_{n\,\pi^*} \leftarrow S_0$ transitions in pyrazine and acridine.

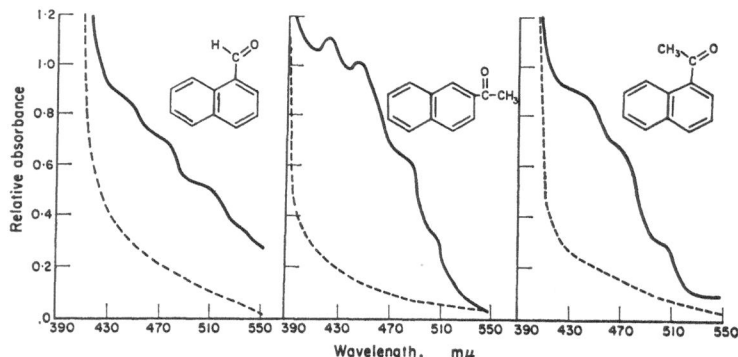

Fig. 23. (- - -) spectrum at atmospheric pressure; (——) spectrum under 100 atm of oxygen. 1-naphthaldehyde, 0.20 M in n-hexane; 2-acetonaphthone, 2.00 M in chloroform; 1-acetonaphthone, 1.00 M in chloroform. Path length 8.5 cm. (From Warwick and Wells, Ref.[56])

A theoretical explanation of the influence of paramagnetic molecules on singlet–triplet intensities involving an exchange interaction between the absorbing molecule and the paramagnetic species was advanced by Hoijtink [57].

Murrell [58] showed that intensity borrowing from charge-transfer states is important in alternant hydrocarbons. Similar conclusions were reached quite generally by Tsubomura and Mulliken [59]: The singlet–triplet transitions derive their intensity from a charge-transfer state, the intensity of the latter resulting mostly from the strong singlet–singlet spectrum of the donor molecule. More recently, Chiu [60] carefully examined the theoretical basis of the exchange-coupling mechanism. The molecule that undergoes the singlet–triplet transition and the neighboring paramagnetic perturber are considered as a single composite molecule. In such a supermolecule, a subsystem can undergo a non-spin-conserving singlet-triplet transition while the overall spin of the composite system remains unchanged.

V. Phosphorescence Spectra

That the existence of a metastable state could be responsible for the long-lived afterglow observed from many aromatic molecules in glassy solvents was suggested by Jablonski [61] in 1935. Six years later Lewis, Lipkin, and Magel [62] were able to determine the absorption spectrum of this metastable or phosphorescent state by irradiating fluorescein in a boric acid glass with a high intensity lamp. They tentatively suggested that the phosphorescent state might be a triplet state, which was definitely established through the now classical work of Terenin [63], Lewis and Kasha [64], and finally through electron paramagnetic resonance by Hutchinson and Mangum [65]. Since then, a vast amount of literature has accumulated. We concentrate here on the description of a high-resolution luminescence spectrometer, an important correlation between the observed phosphorescence lifetime and the triplet-state energy, and finally, some modern techniques of phosphorescence spectroscopy.

1. Measurement of Phosphorescence Spectra

A schematic diagram of a high-resolution luminescence spectrometer for research purposes, recently described by Vo Dinh et al. [66], is shown in Fig. 24.

The light from an intensive Xenon lamp 1 passes the excitation monochromator 2 and excites the sample 3. The emitted sample luminescence is analyzed by the emission monochromator 4 and detected by the photomultiplier 5. Very high sensitivity can be achieved by operating the detector in the single-photon-counting mode. The number of photons observed in a given interval is transferred to a digital recording system 6, enabling further data treatment and data normalization on a large computer [67]. Alternatively, a direct plot of photon intensities can be obtained from the analog recording system 7. The control system 8 determines the scan speed, scan range, and data recording. Emission spectra are obtained by setting the excitation monochromator at a fixed wavelength and scanning the emission monochromator. For excitation spectra, the emission monochromator is set at the wavelength of maximum emission and the excitation monochromator is scanned.

U. P. Wild

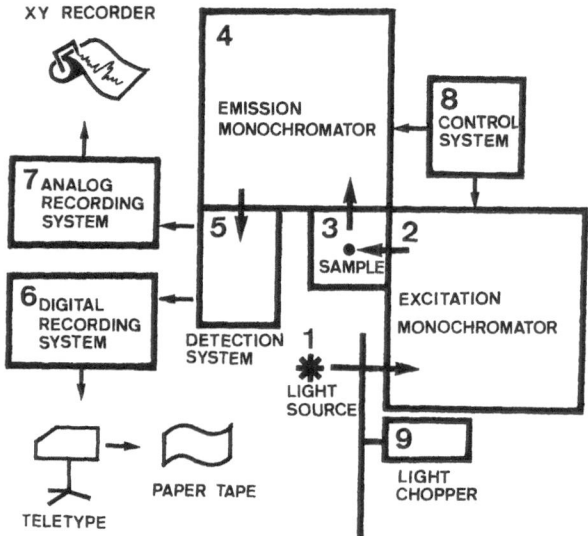

Fig. 24. Schematic diagram of the luminescence spectrometer. (From Vo Dinh and Wild, Ref.[67])

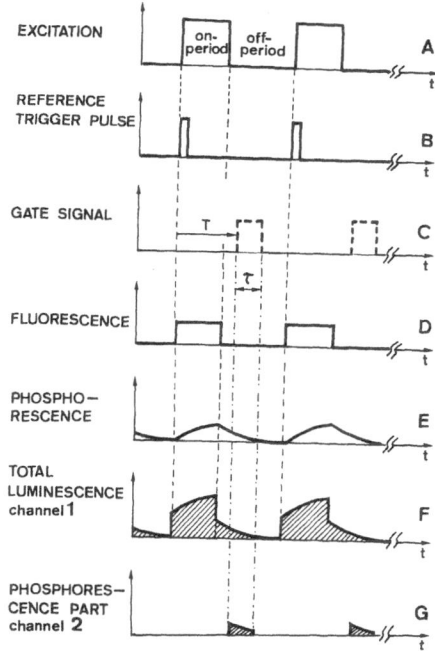

Fig. 25. Simultaneous recording of total luminescence and phosphorescence. (From Vo Dinh and Wild, Ref.[67])

For operational purposes, fluorescence is understood to be short-lived and phosphorescence to be long-lived luminescence. The spectrometer described here uses only a single light chopper 9 and allows simultaneous recording of total luminescence and phosphorescence spectra (Fig. 25).

The sample is excited by chopped irradiation (A). A gate signal (C) of a preselected gate time τ is opened at delay time T after the occurrence of the reference trigger pulse (B) derived from the light chopper. All the photons emitted can be accumulated in channel 1 of a counter giving the total luminescence signal. Recording the number of delayed photons in channel 2 of the counter gives a signal proportional to the phosphorescence intensity. In Fig. 26 the phosphorescence spectrum

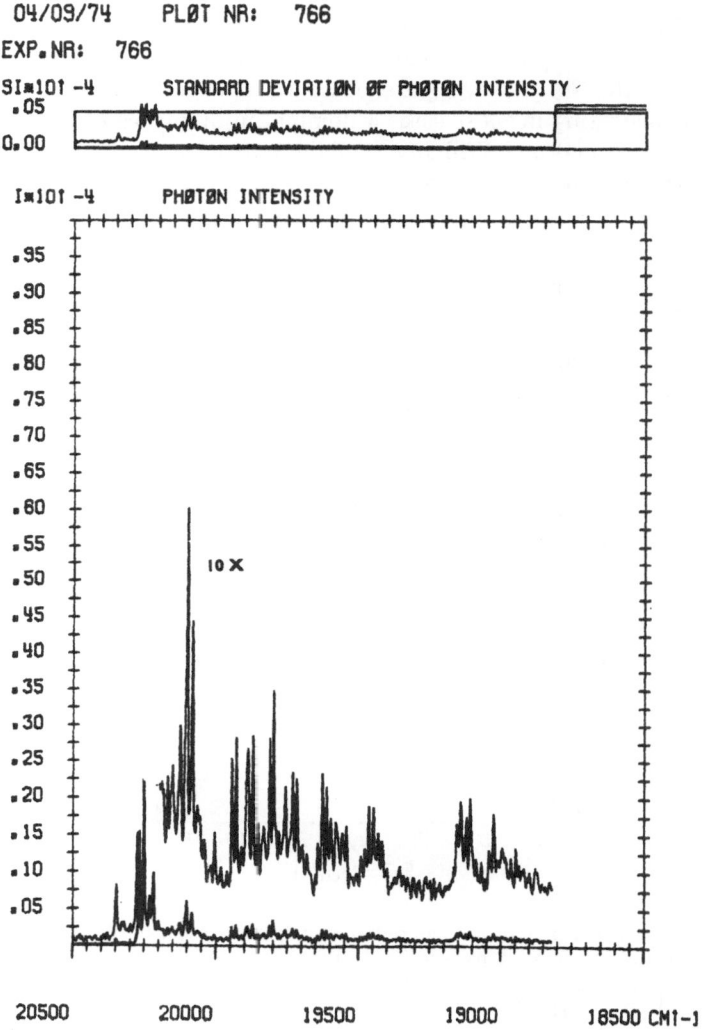

Fig. 26. Phosphorescence spectrum of 2-ethyl-naphthoquinone at 3.6 K in n-heptane. (From Vo Dinh, Holzwarth, Wild, Ref.[68])

of 2-ethyl-naphthoquinone recorded at 3.6 K in n-heptane, a so-called Shpolskii solvent, is shown. Note the abundance of very sharp lines, which results from phononless transitions from the guest molecule occupying different host sites.

2. Phosphorescence Lifetimes

The observed phosphorescence lifetime τ neglecting quenching mechanisms is determined by a radiative deactivation path $(1/\tau_r)$ and by a radiationless one $(1/\tau_{rl})$

$$1/\tau = 1/\tau_r + 1/\tau_{rl} \tag{51}$$

The various mechanisms which affect the $T_1 \leftarrow S_0$ intensity and thus the radiative lifetime have been discussed earlier. For the class of aromatic hydrocarbons spin–orbit coupling is small and a typical value of about 30 sec for τ_r seems appropriate [69]. However, the observed phosphorescence lifetimes vary greatly, demonstrating in most cases a dominating influence of the radiationless contribution. Siebrand and Williams [70] have noticed a very interesting correlation (Fig. 27) between the radiationless deactivation rate $\beta = 1/\tau_{rl}$ and the triplet

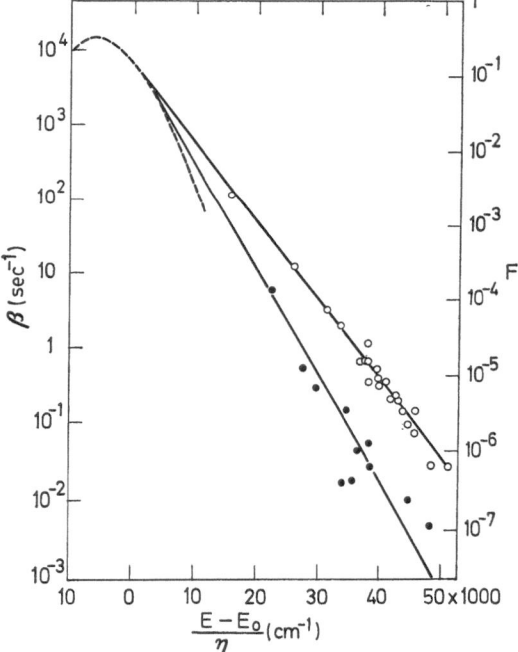

Fig. 27. Semilogarithmic plot of the nonradiative triplet rate constant β against $(E-E_0)/\eta$ for the normal and deuterated hydrocarbons listed in Ref.[71]). The broken line, derived from phosphorescence spectra, is taken from Ref.[71]. The slopes of the two solid lines differ by a factor 1.35. (O:$C_{1-\eta}H_\eta$, $E_0^H = 4000$ cm^{-1}; O:$C_{1-\eta}D_\eta$, $E_0^D = 5500$ cm^{-1}). The following totally deuterated hydrocarbons are included: benzene, triphenylene, acenaphtene, naphthalene, phenanthrene, chrysene, biphenyl, p-terphenyl, pyrene, 1,2-benzanthracene, anthracene (in the order of increasing β). (From Siebrand and Williams, Ref.[70])

energy E of normal and deuterated hydrocarbons: log β is proportional to $(E - E_0)/\eta$, where η is a parameter which describes the relative number of hydrogen or deuterium atoms in the molecule and E_0 ($= 4000$ cm^{-1} for hydrogen and $= 5500$ cm^{-1} for deuterium) represents an average energy accepted by the skeletal modes of the molecule.

The interest aroused by the field of radiationless transitions in recent years has been enormous, and several reviews have been published [72-74]. Basically, the ideas of Robinson and Frosch [75], who used the concepts on non-stationary molecular states and time-dependent perturbation theory to calculate the rate of transitions between Born-Oppenheimer states, are still valid, although they have been extended and refined. The nuclear kinetic energy leads to an interaction between different Born-Oppenheimer states and the rate of radiationless transitions is given by

$$\beta = \frac{2\pi}{\hbar} J^2 \cdot F(E) \cdot \varrho_E \tag{52}$$

where J represents the interaction element between the electronic states, F the Franck-Condon factor, and ϱ_E the density of vibrational states of S_0 at the energy of T_1. The Franck-Condon factor F is obtained from calculating the overlap integrals between vibrational functions

$$F(E) = \prod_k |\langle \varLambda(v_k'')|\varLambda(0_k')\rangle|^2 \tag{53}$$

where the vibrationally excited level of S_0 with an energy $E = \sum_k \hbar\omega_k'' v_k''$ is isoenergetic with the 0-vibrational level of T_1.

The overlap integrals $\langle \varLambda(v_k'')|\varLambda(0_k')\rangle$ decrease sharply with increasing v_k'' values, thus only the normal mode which has the highest frequency, such as the C—H or C—D stretch, will yield small v_k'''s for a given energy E and will be important in radiationless deactivation. The substitution of hydrogen through deuterium lowers the normal modes by approximately $\sqrt{2}$. The corresponding decrease in the efficiency of radiationless deactivation is shown very clearly in Fig. 27.

3. Phosphorescence from Isolated Triplet Sublevels

We have seen in the treatment of spin–orbit coupling that the three sublevels possess very different radiative properties. The generally observed phosphorescence spectrum is a superposition of the three phosphorescence spectra from the sublevels, and it would be very rewarding to single out an individual contribution. The most direct way would seem to use a very high-resolution spectrograph and to resolve the individual phosphorescence lines. Since the width of the optical transitions is large with respect to the zero-field splitting such an experiment is not promising. At very low temperature, under conditions of low spin–lattice relaxation, the triplet levels are "isolated" from each other and emit with their intrinsic lifetimes. In a very interesting experiment Yamauchi and Azumi [76] used time-resolved spectroscopy to sort out the three phosphorescence spectra (Fig. 28) and show that they have rather different vibrational structures. The polarizations of the transitions from the individual zero-field levels of the lowest triplet state in 2,3-dichloroquinoxaline to vibrational levels of the ground state have been discussed and measured in the very thorough work of Tinti and El-Sayed [77]. The group-theoretical predictions involving different coupling mechanisms are summarized in Fig. 29. The "solid" lines gain intensity through direct spin–orbit mixing and are the dominant transitions. The "dashed-dotted" lines can gain in intensity through spin–vibronic coupling; the "dashed" lines require mixing between states of the same electronic type $(^3(\pi,\pi) \longleftrightarrow {}^1(\pi,\pi))$ and are expected to be extremely weak.

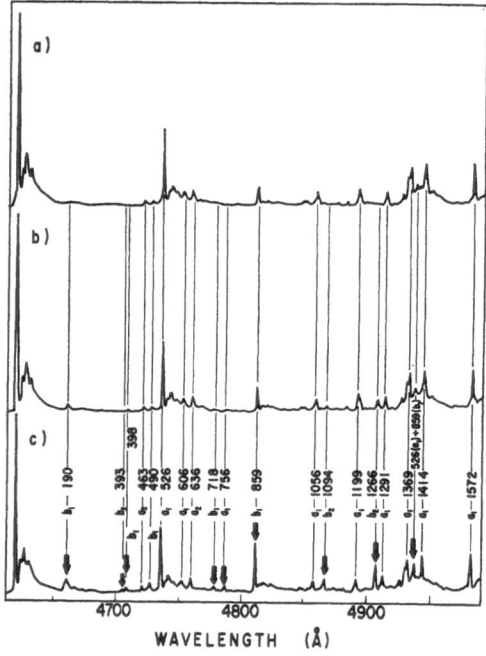

Fig. 28. Time-resolved phosphorescence spectra of quinoxaline in durene host observed at 1.38 K and at (a) 30 msec, (b) 450 msec, and (c) 1500 msec after excitation cutoff. The ordinate scale is normalized with respect to the $0' \to 0''$ band. The numbers shown in (c) represent the vibrational frequencies (in wavenumber unit) measured from the $0' \to 0''$ band (21639 cm^{-1}). The arrows indicate the bands whose relative intensities are remarkably enhanced at later times after the excitation cutoff. (From Yamauchi and Azumi, Ref.[76])

In the experiment done under conditions of inefficient spin–lattice relaxation the triplet spin states originally produced are retained during internal conversion and vibrational relaxation. Important conclusions about the routes of inter-system crossing can thus be obtained from a study of the population of the triplet sublevels, the so-called "spin alignment".

A whole new area of research has been opened by irradiating the triplet system with microwave power and observing its effect on phosphorescence. A very elegant technique for studying dynamics of populating and depopulating the phosphorescent state has been introduced by Schmidt *et al.*[78]. As soon as the phosphorescence of the sublevel with the fastest deactivation rate has decreased sufficiently, a micro-

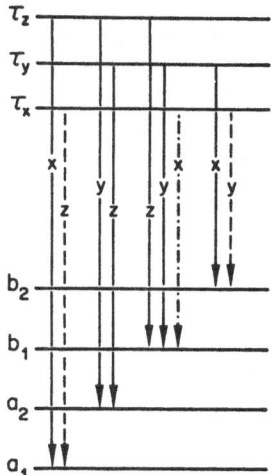

Fig. 29. Group-theoretical predictions of the polarizations of the vibronic transitions, allowed to second order, from the individual zero-field levels of the lowest triplet state of 2,3-dichloro-quinoxaline to vibrational levels of the ground electronic state. "Solid line" transitions gain intensity by spin-orbit mixing between states which differ in the electronic type of one electron (e.g., $S_{n\pi*}$ and $T_{\pi\pi*}$. The "dashed line" transitions require the mixing to occur between states of the same electronic type (e g., $S_{\pi\pi*}$ and $T_{\pi\pi*}$) and is expected to be weaker. The "dash-dotted" transition could involve the favorable mixing between states that differ in the electronic type of one electron, but a spin-vibronic perturbation is needed. (From Tinti and El-Sayed, Ref.[77])

wave sweep is started. The microwave frequency that corresponds to the difference between the strong radiating and one of the "dark" sublevels will connect the two sublevels and introduce a significant repopulation of the radiating level. The strong increase in the "microwave-induced delayed phosphorescence" contains information about the population of the "dark" level and can be clearly seen in Fig. 30.

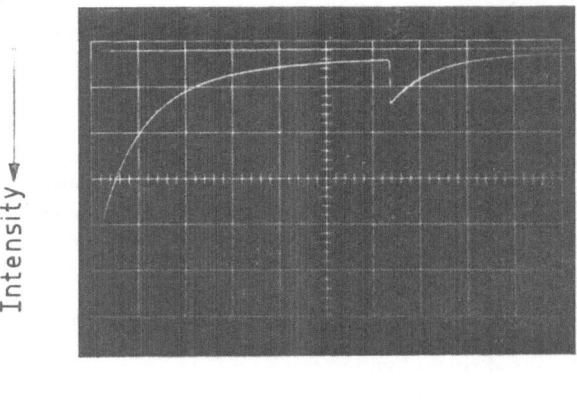

Fig. 30. The decay of phosphorescent quinoline-h_7 in durene at 1.25 K. Abscissa 0.2 sec per division. The delayed signal is induced by sweeping through the $X-Z$ transition at 3585 MHz, 1.28 sec after shutting off the exciting light. (From Schmidt, Antheunis, van der Waals, Ref.[78])

The new techniques of phosphorescence-microwave multiplet resonance spectroscopy with optical detection have been reviewed by El-Sayed [79] and Kwiram [80]. Such exciting experiments as the optical detection on electron–nuclear double resonance (ENDOR) and of electron–electron double resonance (EEDOR) in zero magnetic field have been achieved, and it is certain that much detailed knowledge concerning the phosphorescent states will evolve from this field.

VI. References

[1] Pariser, R., Parr, R. G.: A semi-empirical theory of the electronic spectra and electronic structure of complex unsaturated molecules: I. J. Chem. Phys. *21*, 466—471 (1953).

[2] Pariser, R., Parr, R. G.: A semi-empirical theory of the electronic spectra and electronic structure of complex unsaturated molecules: II. J. Chem. Phys. *21*, 767—776 (1953).

[3] Pople, J. A.: Electron interaction in unsaturated hydrocarbons. Trans. Faraday Soc. *49*, 1375—1385 (1953).

[4] Pople, J. A.: The electronic spectra of aromatic molecules. II: A theoretical treatment of excited states of alternant hydrocarbon molecules based on self-consistent molecular orbitals. Proc. Phys. Soc. (London) *A 68*, 81—89 (1955).

[5] Roothaan, C. C. J.: New developments in molecular orbital theory. Rev. Mod. Phys. *23*, 69—89 (1951).

[6] Mataga, N., Nishimoto, K.: Electronic structure and spectra of some nitrogen heterocycles. Z. Phys. Chem. *12*, 335 (1957); *13*, 140 (1957).

[7] Brillouin, L.: Actualites Sci. Ind. *71* (1933); *159*, (1934).

[8] Herzberg, G.: Electronic spectra of polyatomic molecules, p. 417—419. Princeton, N. J.: Van Nostrand 1966.

[9] Ranalder, U.: Triplet-Triplet Absorptionsspektren von aromatischen Kohlenwasser-stoffen, N-Heterozyklen und Porphyrinen. Dissertation ETHZ (1974).

[10] Zülicke, L.: Quantenchemie, Bd. 1. Grundlagen und allgemeine Methoden, p. 223. Berlin: VEB Deutscher Verlag der Wissenschaften 1973.

[11] Ellis, R. L., Squire, R., Jaffe, H. H.: Use of CNDO method in spectroscopy. V. Spin-orbit coupling. J. Chem. Phys. *55*, 3499 (1971).

[12] Condon, E. U., Shortley, G. H.: Theory of atomic spectra, Chaps. 7 and 8. Cambridge: University Press 1935.

[13] El-Sayed, M. A.: Spin-orbit coupling and the radiationless processes in nitrogen heterocyclics. J. Chem. Phys. *38*, 2834 (1963).

[14] Lower, S. K., El-Sayed, M. A.: The triplet state and molecular electronic processes in organic molecules. Chem. Rev. *66*, 199 (1966).

[15] Förster, Th.: Fluoreszenz Organischer Verbindungen. Göttingen: Vandenhoeck & Ruprecht 1951.

[16] Hameka, H. F.: Advanced quantum chemistry. Theory of interactions between molecules and electromagnetic fields. Reading, Mass.: Addison-Wesley Publishing Company, Inc. 1965.

[17] Hercules, D. M.: Fluorescence and phosphorescence analysis. Principles and applications. New York: Wiley 1966.

[18] Hochstrasser, R. M.: Molecular aspects of symmetry. New York: W. A. Benjamin, Inc. 1966.

[19] The triplet state. Proceesings of an international Symposium held at the American University of Beirut, Lebanon, 14—19 Feb. 1967. Cambridge: University Press 1967.

[20] Bowen, E. J.: Luminescence in chemistry. The Van Nostrand series on physical chemistry. London: D. Van Nostrand Comp. Ltd. 1968.

[21] Goodman, L., Laurenzi, B. J.: Probability of singlet-triplet transitions. In: Advances in quantum chemistry, Vol. 4 (ed. by Per-Olov Löwdin). New York—London: Academic Press 1968.

22) McGlynn, S. P., Azumi, T., Kinoshita, M.: Molecular spectroscopy of the triplet state. Englewood Cliffs, N. J.: Prentice-Hall, Inc. 1969.
23) Becker, R. S.: Theory and interpretation of fluorescence and phosphorescence. New York: Wiley 1969.
24) Birks, J. B.: Photophysics of aromatic molecules. New York: Wiley 1970.
25) Lim, E. C.: Excited states; Vol. I. New York—London: Academic Press, Inc. 1974.
26) Labhart, H., Heinzelmann, W.: Triplet-triplet absorption spectra of organic molecules. In: Organic molecular photophysics. Vol. 1, p. 297. (ed. J. B. Birks). New York: Wiley 1973.
27) Vallotton, M., Wild, U. P.: Design of a flash photolysis apparatus. J. Phys. E. Sci. Instr. *4*, 417 (1971).
28) Wild, U. P.: Aufbau einer Blitzlichtapparatur mit Zündfunkenstrecke und ihre Anwendung zum Studium der Triplett-Triplett Annihilation von Anthracen in Glyzerin. Thesis Nr. 3709 ETHZ (1965).
29) Porter, G., Windsor, M. W.: Triplet states in solution. J. Chem. Phys. *21*, 2088 (1953).
30) Meyer, Y. H., Astier, R., Leclercq, J. M.: Triplet-triplet spectroscopy of polyacenes. J. Chem. Phys. *56*, 801 (1972).
31) Labhart, H.: Eine experimentelle Methode zur Ermittlung der Singlett-Triplett-Konversionswahrscheinlichkeit der Triplettspektren von gelösten organ. Molekülen: Messungen an 1,2-Benzanthracen. Helv. Chim. Acta *47*, 2279 (1964).
32) Hunziker, H. E.: Gas-phase absorption spectrum of triplet naphthalene in the 220—300-nm and 410—620-nm wavelength regions. J. Chem. Phys. *56*, 400 (1972).
33) Craig, D. P., Ross, I. G.: The triplet-triplet absorption spectra of some aromatic hydrocarbons and related substances. J. Chem. Soc. *1954*, 1589 (1954).
34) Brinen, J. S.: Application of electron spin resonance in the study of triplet states. III. Extinction coefficients of triplet-triplet transitions. J. Chem. Phys. *49*, 586 (1968).
35) Ranalder, U. B., Känzig, H., Wild, U. P.: The determination of molar absorption coefficients of metastable states. J. Photochem. *4*, 97 (1975).
36) El-Sayed, M. A., Pavlopoulos, T.: Polarization of the triplet-triplet spectrum of some polyacenes by the method of photoselection. J. Chem. Phys. *39*, 834 (1963).
37) Pavlopoulos, T. G.: Search for the first $^3A_{1g}^- \leftarrow {}^3B_{2u}^+$ transition in naphthalene. J. Chem. Phys. *53*, 4230 (1970).
38) Orloff, M. K.: Theoretical study of triplet-triplet absorption spectra. I. Alternant hydrocarbon molecules. J. Chem. Phys. *47*, 235 (1967).
39) Lavalette, D.: Polarized excitation spectrum of the triplet-triplet absorption of aromatic hydrocarbons. Chem. Phys. Letters *3*, 67 (*1969*).
40) Moffitt, W.: The electronic spectra of cata-condensed hydrocarbons. J. Chem. Phys. *22*, 320 (1954).
41) Kearns, D. R.: Determination of the assignment of triplet states in cata-condensed aromatic hydrocarbons. J. Chem. Phys. *36*, 1608 (1962).
42) Nouchi, G.: Etude et interprétation des spectres d'absorption triplet-triplet. Classification des états triplets supérieurs. II. Les hydrocarbures aromatiques. J. Chim. Phys. *66*, 554 (1969).
43) de Groot, R. L., Hoijtink, G. J.: Triplet-triplet transitions in naphthalene. J. Chem. Phys. *46*, 4523 (1967).
44) Hoijtink, G. J.: Triplet-triplet spectra of alternant hydrocarbon molecules. Pure Appl. Chem. *11*, 393 (1965).
45) Pancir, J., Zahradnik, R.: Theoretical study of singlet-triplet and triplet-triplet spectra. I. Selection of parameters and the basis of configuration interaction in closed shell and restricted open shell semiempirical methods. J. Phys. Chem. *77*, 107 (1973).
46) Pancir, J., Zahradnik, R.: Theoretical study of singlet-triplet and triplet-triplet spectra. II. Conjugated hydrocarbons. J. Phys. Chem. *77*, 114 (1973).
47) Rothman, W., Case, A., Kearns, D. R.: Determination of singlet-triplet absorption spectra from phosphorescence excitation spectra: α-bromonaphthalene. J. Chem. Phys. *43*, 1067 (1965).
48) Kearns, D. R., Case, W. A.: Investigation of singlet-triplet transitions by the phosphorescence excitation method. III. aromatic ketones and aldehydes. J. Am. Chem. Soc. *88*:22, 5087 (1966).

49) Jones, C. R., Kearns, D. R., Wing, R. M.: Investigation of singlet-triplet transitions by phosphorescence excitation spectroscopy. X. A simple π,π^*-unsaturated ketone. J. Chem. Phys. *58*, 1370 (1973).

50) Marchetti, A. P., Kearns, D. R.: Investigation of singlet-triplet transitions by the phosphorescence excitation method. IV. The singlet-triplet absorption spectra of aromatic hydrocarbons. J. Am. Chem. Soc. *89*, 768 (1967).

51) Hirota, N.: Use of triplet-state energy transfer in obtaining singlet-triplet absorption in organic crystals. J. Chem. Phys. *44*, 2199 (1966).

52) Avakian, P., Abramson, E.: Singlet-triplet excitation spectra in naphthalene and pyrene crystals. J. Chem. Phys. *43*, 821 (1965).

53) Evans, D. F.: Perturbation of singlet-triplet transitions of aromatic molecules by oxygen under pressure. J. Chem. Soc. *1957*, 1351.

54) Evans, D. F.: Magnetic perturbation of singlet-triplet transitions. Part III: Benzene derivatives and hererocyclic compounds. J. Chem. Soc. *1959*, 2753.

55) Evans, D. F.: Magnetic perturbation of singlet-triplet transitions. Part V: Mechanism. J. Chem. Soc. *1961*, 1987.

56) Warwick, D. A., Wells, C. H. J.: Perturbation of singlet-triplet transitions in aromatic carbonyl compounds. Spectrochim. Acta *24 A*, 589 (1968).

57) Hoijtink, G. J.: The influence of paramagnetic molecules on singlet-triplet transitions. Mol. Phys. *3*, 67 (1960).

58) Murrell, J. N.: The effect of paramagnetic molecules on the intensity of spinforbidden absorption bands of aromatic molecules. Mol. Phys. *3*, 319 (1960).

59) Tsubomura, H., Mulliken, R. S.: Molecular complexes and their spectra. XII. Ultraviolet absorption spectra caused by the interaction of oxygen with organic molecules. J. Am. Chem. Soc. *82*, 5966 (1960).

60) Chiu, Y. N.: On singlet-triplet transitions induced by exchange with paramagnetic molecules and the intermolecular coupling of spin angular momenta. J. Chem. Phys. *56*, 4882 (1972).

61) Jablonski, A.: Über den Mechanismus der Photolumineszenz von Farbstoffphosphoren. Z. Physik *94*, 38 (1935).

62) Lewis, G. N., Lipkin, D., Magel, T. T.: Reversible photochemical processes in rigid media. A study of the phospLorescent state. J. Am. Chem. Soc. *63*, 3005 (1941).

63) Terenin, A.: Zh. Fiz. Khim. *18*, 1 (1944).

64) Lewis, G. N., Kasha, M.: Phosphorescence and the triplet state. J. Am. Chem. Soc. *66*, 2100 (1944).

65) Hutchison, C. A., Mangum, B. W.: Paramagnetic resonance absorption in naphthalene in its phosphorescent state. J. Chem. Phys. *34*, 908 (1961).

66) Vo Dinh, T., Wild, U. P.: High resolution luminescence spectrometer. 1: Simultaneous recording of total luminescence and phosphorescence. Appl. Opt. *12*, 1286 (1973).

67) Vo Dinh, T., Wild, U. P.: High resolution luminescence spectrometer. 2: Data treatment and corrected spectra. Appl. Opt. (Dec. 1974).

68) Vo Dinh, T., Holzwarth, A., Wild, U. P.: To be published.

69) Kellogg, R. E., Bennett, R. G.: Radiationless intermolecular energy transfers. III. Determination of phosphorescence efficiencies. J. Chem. Phys. *41*, 3042 (1964).

70) Siebrand, W., Williams, D. F.: Isotope rule for radiationless transitions with an application to triplet decay in aromatic hydrocarbons. J. Chem. Phys. *46*, 403 (1967).

71) Siebrand, W.: Mechanism of radiationless triplet decay in aromatic hydrocarbons and the magnitude of the Franck-Condon factors. J. Chem. Phys. *44*, 4055 (1966).

72) Henry, B. R., Siebrand, W.: Radiationless transitions. In: Organic molecular photophysics, Vol. 1, p. 153. New York: Wiley 1973.

73) Jortner, J., Rice, S. A., Hochstrasser, R. M.: Advances in photochemistry, Vol. 7 (eds. Noyes, Pitts, and Hammond). New York: Wiley 1969.

74) Freed, K. F.: The theory of radiationless processes in polyatomic molecules. In: Topics Curr. Chem. *31*, 105 (1972).

75) Robinson, G. W., Frosch, R. P.: Electronic excitation transfer and relaxation. J. Chem. Phys. *38*, 1187 (1963).

76) Yamauchi, S., Azumi, T.: Observation of the phosphorescence spectra from the spin sublevels of low emissivity: Quinoxaline. Chem. Phys. Letters *21*, 603 (1973).

77) Tinti, D. S., El-Sayed, M. A.: New techniques in triplet state phosphorescence spectroscopy: Application to the emission of 2,3-dichloroquinoxaline. J. Chem. Phys. *54*, 2429 (1971).

78) Schmidt, J., Antheunis, D. A., van der Waals, J. H.: The dynamics of populating and depopulating the phosphorescent triplet state as studied by microwave induced delayed phosphorescence. Mol. Phys. *22*, 1 (1971).

79) El-Sayed, M. A.: Phosphorescence-microwave multiple resonance spectroscopy. In: MTP International Review of Science, Physical Chemistry Series One, Vol. 3.

80) Kwiram, A. L.: Optical detection of magnetic resonance in molecular triplet states. In: MTP International Review of Science, Physical Chemistry Series One, Vol. 4.

Received September 10, 1974

Reactions of Aromatic Nitro Compounds *via* Excited Triplet States

Prof. Dr. Dietrich Döpp

Fachbereich Chemie der Universität Trier-Kaiserslautern in Kaiserslautern

Contents

Introduction 51

A. Hydrogen Abstractions, Photoreduction and Related Reactions........ 51

    I. Nitrobenzene and its Derivatives............................. 51

        1. Photoreduction of Nitrobenzene............................ 51

        2. Photoreduction of Substituted Nitrobenzenes............... 55

        3. Incorporation of Solvent Fragments....................... 57

        4. Intramolecular Hydrogen Abstractions..................... 58

        5. Addition of Nitrobenzenes to π-Systems................... 61

    II. Photoreduction of Nitronaphthalenes and Nitrobiphenyls........ 63

        1. Spectroscopy of the Nitronaphthalenes..................... 63

        2. Photoreduction of 1-Nitronaphthalene...................... 64

        3. Photoreduction of 2-Nitronaphthalene...................... 65

        4. Photoreduction of Nitrobiphenyls.......................... 65

    III. Photoreduction of Nitropyridines and Nitropyridine-N-oxides..... 66

        1. 4-Nitropyridine 66

        2. 4-Nitropyridine-N-oxides 67

B. Nucleophilic Aromatic Photosubstitutions 68

    I. General Remarks 68

    II. Orientation Rules and Representative Examples................ 69

    III. Details of Nucleophilic Aromatic Photosubstitutions............ 72

        1. Nitrobenzene Derivatives................................. 72

        2. Nitronaphthalenes, Nitroazulenes and Nitrobiphenyls......... 75

        3. Nitrosubstituted Heterocycles............................. 77

Introduction

Interest in the photochemistry of nitro compounds has increased considerably during the last decade. Progress has been made in two major fields, namely

(1) photoreduction and related reactions,

(2) nucleophilic aromatic photosubstitution.

The former mode of reduction is widely regarded as typical for (n,π^*), the latter for (π,π^*)-excited states.

Spectroscopic data of nitroaromatics have been reviewed [1]; in addition, several papers [2-5] on luminescence of nitroaromatic compounds have appeared recently. The phosphorescence polarization of several aromatic nitro compounds has been studied [5] and recent triplet–triplet absorption data on 1- and 2-nitronaphthalene have become available [6].

The spectroscopy [7] and photochemistry [8] of nitrocompounds in general have also been reviewed previously as well as theoretical aspects of the C–NO_2 bond [9].

Although many light-induced reactions of nitro compounds are known, attempts to elucidate the multiplicity of the reacting excited state have been made in only a fraction of cases. Conclusive information is not easily obtained since in general nitroaromatics suffer from the disadvantage that they show very weak or even no fluorescence *and* have very short triplet lifetimes.

Short triplet lifetimes, moreover, cause difficulties in applying conventional flash photolysis techniques, as will be shown later, and demand nanosecond flash methods.

Indirect methods, such as sensitization, quenching, or product pattern studies, have been widely used. Although these methods are by no means a good substitute for spectroscopic techniques and suffer from serious drawbacks, one should not be discouraged from investigating the photochemistry of nitro compounds.

A comprehensive treatment of nitro-compound photochemistry is not attempted here and would be beyond the scope of this volume. The discussion is thus restricted to aromatic and heteroaromatic nitro compounds. The reactions preferentially reviewed are those for which at least some experimental information has been given on efforts to elucidate the multiplicity of the reacting excited state, or those in which intermediate excited triplet states seem highly probable for various reasons.

A. Hydrogen Abstractions, Photoreduction and Related Reactions

I. Nitrobenzene and its Derivatives

1. Photoreduction of Nitrobenzene

The photoreduction of nitrobenzene had already been observed by Ciamician and Silber [10]. More recently, quantitative studies have been carried out by Testa and his group [11-15] and other authors [16-18]. It is generally accepted [11-15, 17,18] that the long wavelength ($\sim$340 nm) absorption of nitrobenzene must be assigned [19,20] to a n → π^*-transition. Promotion of an electron out of a lone pair of oxygen into the π-system imparts radical character to the oxygens and is responsible for the electrophilic nature of the (n,π^*)-excited nitro group.

The photolysis (366 nm) of nitrobenzene in 2-propanol leads to phenyl hydroxylamine as the first stable and identifiable product [11]. The initial stages of the reaction are conveniently depicted as follows [11]:

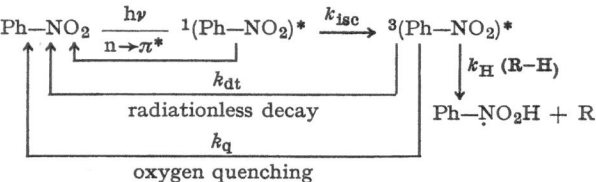

Several subsequent dark reactions involving Ph–$\overset{\bullet}{N}O_2H$ and R· have been discussed elsewhere (and are not repeated here), to account for both the formation of phenylhydroxylamine and acetone [8,11,16,18] and to explain the nature of radicals detected in ESR-Studies upon photolysis of nitrobenzenes in hydrogen donor solvents [21–24].

The paramagnetic species observed is not the elusive hydroxyphenylaminyloxide (A) as postulated in earlier work [25] but an alkoxyphenylaminyloxide (B) formed by addition of R· to a molecule of ground state nitrobenzene:

$$
\begin{array}{lll}
\overset{\displaystyle O\cdot}{\underset{\displaystyle |}{}} & \overset{\displaystyle O\cdot}{\underset{\displaystyle |}{}} & \\
Ph{-}N{-}OH \quad (A) & Ph{-}N{-}OR \quad (B) & R = \text{—} , C(CH_3)_2OH.
\end{array}
$$

The assignment [26] of Ph–NH–O· to the radical observed upon irradiation of nitrobenzene in 2-propanol has been questioned [24].

The evidence for triplet multiplicity of the reacting excited state is as follows:

The generation of a paramagnetic species by irradiation of a 10^{-2} M solution of nitrobenzene in thoroughly degassed tetrahydrofuran at room temperature is efficiently quenched [27] by 10^{-1} mole l^{-1} perfluornaphthalene ($E_T = 56.6$ kcal mole^{-1}).

Details of nitrobenzene photochemistry reported by Testa are consistent with the proposal that the lowest triplet excited state is the reactive species. Photoreduction, as measured by disappearance quantum yields of nitrobenzene in 2-propanol is not very efficient: $\phi = (1.14 \pm 0.08) \cdot 10^{-2}$ [11]. On the other hand, the triplet yield of nitro benzene in benzene, as determined by the triplet–counting method of Lamola and Hammond [28] is 0.67 ± 0.10 [13]. This raises the question of the cause of inefficiency in photoreduction. Whereas Lewis and Kasha [29] report the observation of nitrobenzene phosphorescence, no long–lived emission from carefully purified nitrobenzene could be detected by other authors [14,30]. Unfortunately, the literature value of E_T for nitrobenzene (60 kcal mole^{-1}) is thus based on an impurity emission and at best a value between 60 and 66 kcal mole^{-1} can be envisaged from energy–transfer experiments [30].

Low phosphorescence efficiency, however, leaves fast radiationless decay as the prime course of inefficient photochemistry.

From $k_H = 0.8 \times 10^6$ 1 mole^{-1} s^{-1} [14] and straightforward kinetics the rate constant for radiationless decay of triplet nitrobenzene could be derived: $k_{dt} = 0.6 \times 10^9$ s^{-1}. Thus it is easily unterstood why oxygen quenching (assumed to be diffusion controlled [31]) does not affect photoreduction in 2-propanol to a large extent: upon irradiation in air–saturated ($[O_2] \approx 10^{-3}$ M) solutions, the quantum yield of nitrobenzene disappearance is lowered only by 30% and $k_q \times [O_2] \approx 10^6$ s$^{-1} \ll k_{dt}$.

A short lifetime ($\tau \approx 10^{-9}$ s) for the lowest excited triplet of nitrobenzene has also been obtained from energy–transfer experiments using cis-piperylene as quencher. Even at concentrations of 2.4 mole 1^{-1} *cis*-piperylene not all nitrobenzene triplets were quenched [13], but singlet energy–transfer has been disfavored by kinetic reasoning [13].

With good hydrogen donors photoreduction of nitrobenzene becomes more efficient: the rate constant for hydrogen abstraction from tributylstannane by $^3(n, \pi^*)$-nitrobenzene has been determined as 4×10^8 1 mole^{-1}s^{-1} [14].

Nitrobenzene may be used to oxidize methylene groups photochemically [32]. Irradiation in cyclohexane results in the formation of cyclohexanol and cyclohexanone [32]. With 2-methylbutane, selectivities of 1:19:300 have been observed for attack on primary, secondary and tertiary hydrogens respectively, using pyrex filtered light. With a vycor filter, the corresponding figures were 1:7:110 [32]. The potential use of nitrobenzene moieties attached to a steroid skeleton for dehydrogenation at remote positions has been investigated [33].

The efficiency of nitrobenzene photoreduction may be increased remarkably in 2-propanol/hydrochloric acid mixtures. In 50% 2-propanol/water containing 6 moles 1^{-1} HCl, acetone and a complex mixture of chlorinated reduction products are formed [18]. Both HCl and 2-propanol (as hydrogen source) are needed. When sulfuric acid is substituted for HCl, enhanced photoreduction does not occur. When using mixtures of HCl and LiCl to maintain a constant chloride concentration (6 M) and vary [H$^+$], a constant disappearance quantum yield $\phi_{366} = 0.15$ is found within the [H$^+$]-range 0.05—6 moles 1^{-1}. This strongly suggests that chloride ions play an essential role, probably *via* electron transfer to $^3(n, \pi^*)$-nitrobenzene [18] [Eq. (1)], but it is also evident from the data presented [12], that the presence of acid is probably important in subsequent steps, [Eq. (3)].

+ 3,7% Ar—N=N—Ar + 12,5% Ar—N(O)=N—Ar

(Ar = 4—Cl—C$_6$H$_4$)

$$(Ph-NO_2)^* + Cl^\ominus \longrightarrow [Ph-N\dot{O}_2^\ominus, Cl\cdot] \qquad (1)$$

$$[Ph-N\dot{O}_2^\ominus, Cl\cdot] \longrightarrow Ph-NO_2 + Cl^\ominus \qquad (2)$$

$$[Ph-N\dot{O}_2^\ominus, Cl\cdot] + H^\oplus \longrightarrow [Ph-\dot{N}O_2H, Cl\cdot] \qquad (3)$$

$$[Ph-\dot{N}O_2H, Cl\cdot] \longrightarrow Ph-NO_2 + H^\oplus + Cl^\ominus \qquad (4)$$

$$[Ph-\dot{N}O_2H, Cl\cdot] + (CH_3)_2CHOH \longrightarrow Ph-\dot{N}O_2H + HCl + (H_3C)_2\dot{C}OH \qquad (5)$$

$$(CH_3)_2\dot{C}OH + Ph-NO_2 \longrightarrow (CH_3)_2CO + Ph-\dot{N}O_2H \qquad (6)$$

$$2\ Ph-\dot{N}O_2H \longrightarrow Ph-NO_2 + Ph-N(OH)_2 \qquad (7)$$

$$Ph-N(OH)_2 \longrightarrow Ph-NO + H_2O \qquad (8)$$

This scheme [18] implies formation of one mole of acetone per mole of nitrobenzene being consumed, which agrees with the experimental results. Nitrosobenzene is an attractive intermediate, since it had been shown independently to undergo chlorination/reduction to the product pattern given above in a dark reaction [18,34].

Electron transfer [Eq. (1)] would occur at a rate near the diffusion limit if it were exothermic. However, a close estimate of the energetics including solvation effects has not been made yet. Recent support of the intermediacy of a charge transfer complex such as $[Ph-N\dot{O}_2^\ominus, Cl\cdot]$ comes from the observation of a transient ($\lambda_{max} \approx 440$ nm, $\tau = 2.7 \pm 0.5$ ms) upon flashing (80 J, 40 µs pulse) a degassed solution (50% 2-propanol in water, 4×10^{-4} M in nitrobenzene, 6 moles l^{-1} HCl)[15]. The absorption spectrum of the transient is in satisfactory agreement with that of $Ph-\dot{N}O_2H$, which in turn arises from rapid protonation of $Ph-NO_2^\ominus$ under the reaction conditions:

$$Ph-\dot{N}O_2H \rightleftarrows Ph-N\dot{O}_2^\ominus + H^\oplus \ (pK_a = 3.2)\ .$$

It should be mentioned that irradiation of nitrobenzene in aqueous (no alcohol added) hydrochloric acid at room temperature also yields 44–62% 2,4,6-trichloro- and 10% 2,4-dichloroaniline in an undoubtedly complicated reaction [35], very likely also initiated by electron transfer [Eq. (1)].

Protonation of $^3(n, \pi^*)$-nitrobenzene had been suggested earlier [12] and later questioned [18] on account of an estimated extremely weak basicity of $^3(n, \pi^*)$-nitrobenzene. Enhanced basicity of the lowest excited singlet state compared to ground and lowest excited triplet state has been derived from shifts in the phosphorescence and absorption spectra of nitrophenols [36]. On this basis, the increased rate of nitrobenzene photoreduction in acidic solution is found to be thermodynamically unfeasible in the lowest excited triplet state [36]. Although it might be thermodynamically feasible in the excited singlet state, the short lifetime of the latter state may make this possibility unlikely.

The reduction of nitrobenzene may also be accomplished by triplet excited pyrochlorophyll [37]. Comparison of the quantum yields of such photoreductions by hydrazobenzene in ethanol-pyridine solutions with the polarographic quarter wave potential of the nitro compounds leads to the conclusion that electron transfer can occur from triplet excited pyrochlorophyll to the nitro compounds without input of additional activation energy, provided that the potential of the nitro compound is greater than -0.80 V in the solvent described [37].

The photoreduction of nitrobenzene using pyrex filtered light from a medium pressure mercury arc was studied in petroleum, toluene, ether, 2-propanol, *tert*-butyl alcohol, diethylamine, triethylamine, aqueous solutions of 2-propanol and diethylamine [16] and also in aqueous t-butylalcohol containing sodium borohydride [38]. Varying amounts of aniline, azo- and azoxybenzene were obtained. In the presence of a fourty-fold excess of benzophenone, a six-fold increase in the rate of aniline formation in ethereal solution was observed, and aniline formation was completely suppressed by addition of biacetyl or octafluornaphthalene [16]. Since unreacted nitrobenzene could be recovered in these experiments, it is demonstrated that the triplet state of nitrobenzene was quenched.

It should be noted at this point that the mechanism of photoreduction in amine solvents is highly likely to be quite different from that in hydrogen donor solvents. In the former class of solvents, electron transfer seems to prevail. Sterically hindered nitrobenzenes are not capable of hydrogen abstraction from hydrogen donors as ethers or 2-propanol (see Section A. I. 4) but are efficiently photoreduced in di- or triethylamine [39].

Photochemical deoxygenation of nitrobenzene to nitrosobenzene with cyanide ions [38,40] or by molecular complexation [41] with boron trichloride have been reported. No experiments to elucidate the multiplicity of the reacting excited state have been described, however.

2. Photoreduction of Substituted Nitrobenzenes

The photoreduction of eight electron-acceptor- and four electron-donor-substituted nitrobenzenes has been studied and quantum yields for either starting material disappearance or product formation have been reported [17]. Photolysis of 4-nitrobenzonitrile and 4-nitrotoluene in air–saturated solutions was completely quenched and thus a triplet multiplicity of the reacting excited state was derived[17].

Nitromesitylene and 2,6-dichloronitrobenzene were photoreduced in 2-propanol [42]. 3- and 4-nitroanisole, 3-nitrototuene and 2,6-dimethylnitrobenzene were efficiently photoreduced in ether or in aliphatic amines mainly to the corresponding anilines [16]. No experiments to prove the intermediacy of triplet states have been reported in these cases.

Various substituted nitrobenzenes have been irradiated in

a) alkaline aqueous methanol [43a, b] and
b) aqueous-methanolic formate buffers [43b] (with formate ion regarded as hydride donor)

to yield mainly the corresponding anilines. Enhanced rates of photoreduction, compared to parallel experiments with no alkali or formate ion present, were observed. For 1,3-dinitrobenzene and 4,4'-dinitrobiphenyl on irradiation in methanolic methoxide solution, straight line plots have been obtained for $\phi_{\text{dissap}}^{-1}$ vs. $[\text{OCH}_3^{\ominus}]^{-1}$, which is in agreement with a bimolecular reaction between excited starting material and methoxide and has been interpreted in terms of photoreduction *via* electron-transfer at least in basic media [43 b]. The authors did not observe chargetransfer absorptions. No comment on photosubstitution by methoxide as another cause for starting material disappearance has been made.

Photoreduction of m-dinitrobenzene is suppressed by nitric oxide and atmospheric oxygen [43 a]. Quenching by naphthalene, which also has been observed, does not necessarily imply a triplet excited reacting state, since quenching of the excited singlet by naphthalene would be an alternative possibility.

A general scheme for electron transfer from methoxide, similar to that of electron transfer from chloride [18], is as follows [43 b]:

$$Ar-NO_2 \xrightarrow{h\nu} {}^1(Ar-NO_2)^* \longrightarrow {}^3(Ar-NO_2)^*$$

$$(CH_3O^\ominus) \text{ quenching}$$

$$\downarrow CH_3O^\ominus (e^\ominus \text{ transfer})$$

$$(Ar-NO_2H)^\ominus + CH_2O \xleftarrow[\substack{\text{internal} \\ \text{H-transfer}}]{} [Ar-N\dot{O}_2^\ominus \cdots\cdots CH_3O\cdot]$$

$$\downarrow$$

$$ArN\dot{O}_2^\ominus + CH_3O\cdot$$

$$(Ar-NO_2H)^\ominus \xrightarrow{H^\oplus} ArN(OH)_2 \longrightarrow Ar-NO + H_2O \text{ etc.}$$

Correlations between the energy of the lowest excited triplet state (E_T) and the one electron reduction potentials $(-\varepsilon_{1/2})$ at a dropping mercury electrode have been obtained for nineteen aromatic nitro compounds [44]. The existence of three distinct E_T vs. $(-\varepsilon_{1/2})$-relationships, corresponding to three classes of excited state reactivities, has been demonstrated. Among the first two groups (I and II) of compounds, plots of E_T vs. first wave $-\varepsilon_{1/2}$ gave straight lines, whereas no linear relationship could been observed for group III.

Group I comprises nitroaromatics with lowest (n, π*)- [or (π,π*)- with considerable (n, π*)-contribution] excited states, the triplets of which behave electrophilically and tend to abstract hydrogen atoms or electrons and thus typically undergo photoreduction. Compounds which fit this correlation were: Nitrobenzene; 2-nitro-, 4-nitro- and 4,4'-dinitrobiphenyl, 2-nitrofluorene, 3-nitroacetophenone, 1,8- and 1,5-dinitronaphthalene.

Group II compounds may be reduced or may undergo photosubstitution of the nitro group under conditions which favour photoreduction of group I compounds. Charge transfer contributions to the lower excited states decrease the electrophilicity of the excited nitro group. Examples given were: 1-nitro-, 2-nitro- and 1,4-dinitronaphthalene; 8-nitroquinoline. 4-Nitroanisole (E_T 60.8 kcal mole^{-1} [45], 59.5 kcal mole^{-1} [4]) does not fit into this correlation, although it is prone to undergo photosubstitution (see Section B).

Nitroanilines and nitronaphthylamines (group III) show remarkable stability towards both photoreduction and photosubstitution. Photoreduction (including intramolecular hydrogen abstraction) occurs, however, after acylation of the amino group in nitroanilines [46-48]. The stability of nitroanilines towards photoreduction is evidently due to the charge transfer character of the lowest excited state. Another possibility could be the increased probability of radiationless deactivation due to smaller separation of the ground state and the lowest excited states.

3. Incorporation of Solvent Fragments

Few cases of incorporation of solvent fragments following hydrogen abstraction have become known.

Photoreduction of N-benzoyl-4-nitroaniline in primary alcohols gives rise to 4-acylaminophenylbenzamides [47]:

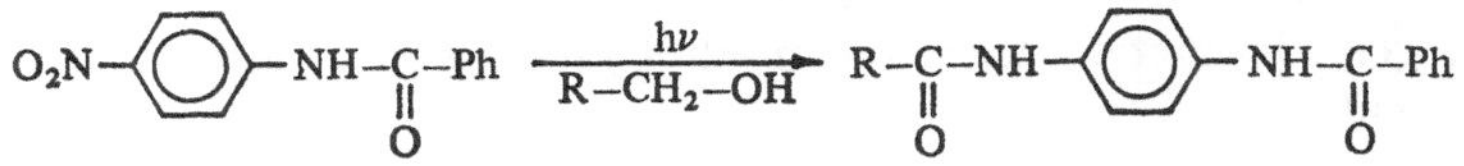

Photolysis of dinitroprehnitene (*1*) in ether yields 25% 2-nitro-3,4,5,6-tetramethylaniline (*i. e.* the normal reduction product) and 5% of the imine *4*, which is probably formed as outlined below [50a]:

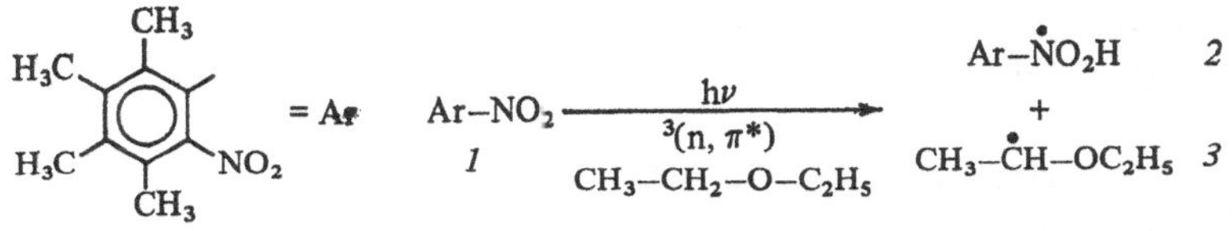

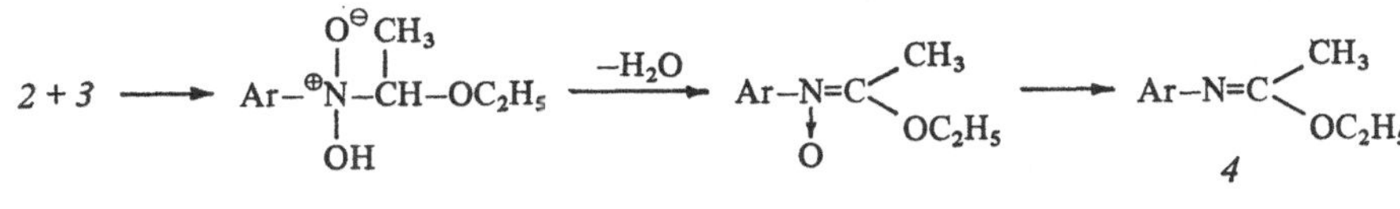

An example of radical coupling following hydrogen abstraction by excited nitroethane from cyclohexane or diethyl ether in solution has also been reported [50b].

Formation of α-methyl-N-arylnitrones is observed during photoreduction (*via* electron transfer) of sterically hindered nitrobenzenes in triethylamine [39]:

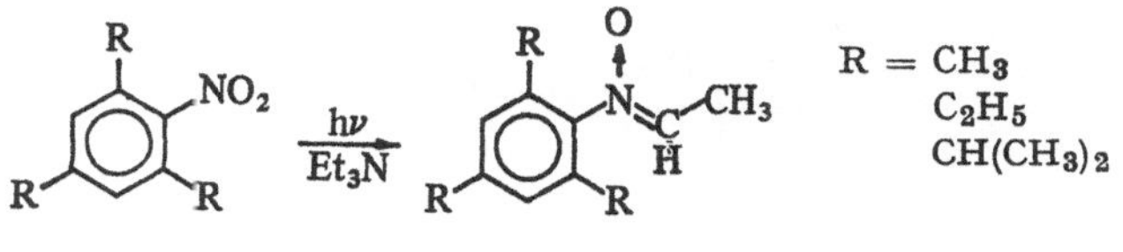

A rationale may be given as follows:

$$Ar-NO_2^* + (C_2H_5)_3N \longrightarrow [Ar-NO_2^{\ominus} + (C_2H_5)_3\overset{\cdot}{N}{}^{\oplus}] \longrightarrow$$

$$\longrightarrow [Ar-\overset{\cdot}{N}O_2H + (C_2H_5)_2N-\overset{\cdot}{C}H-CH_3] \longrightarrow Ar-N(OH)-O-CHCH_3-N(C_2H_5)_2$$

$$\longrightarrow Ar-N=O + CH_3-CHOH-N(C_2H_5)_2$$

$$Ar-N=O \longrightarrow \longrightarrow Ar-NHOH$$

$$Ar-NHOH + CH_3-CHOH-N(C_2H_5)_2 \longrightarrow Ar-\overset{\overset{\textstyle O}{\uparrow}}{N}=CH-CH_3 + (C_2H_5)_2NH + H_2O$$

$$(Ar = 2,4,6-R_3-C_6H_2, \; R = CH_3, \; C_2H_5, \; or \; CH(CH_3)_2)$$

The multiplicity of the reacting excited state has not yet been determined.

A different mode of reaction, however, is observed in photoreductions of nitroaromatics by aromatic tertiary amines. Irradiation of benzene solutions of N-methylated anilines and either m-chloronitrobenzene or 1-nitronaphthalene results in oxidative demethylation of the amines accompanied with reduction of the nitro compound to the corresponding arylamine [49]. The authors suggest that hydrogen abstraction from the methyl group takes place as the primary chemical event.

$$Ar-NO_2 + Ph-NRCH_3 \xrightarrow{h\nu} Ar-NH_2 + Ph-NHR + Ph-NR-CHO$$

$$(Ar = 3-Cl-C_6H_4 \; or \; \alpha\text{-naphthyl}, \; R = H \; or \; CH_3).$$

4. Intramolecular Hydrogen Abstractions

For a detailed review of the photochemistry of 2-nitrobenzaldehyde and its derivatives as well as of o-nitrobenzyl compounds in general the reader is referred to Ref. [8]. Photochromic systems of the o-nitrobenzyl type will also not be discussed here.

It has been reported recently, that the transformations $1 \rightarrow 2$ and $3 \rightarrow 4$ are not markedly influenced by dissolved oxygen under atmospheric pressure and thus a singlet excited state was proposed as reactive species [42].

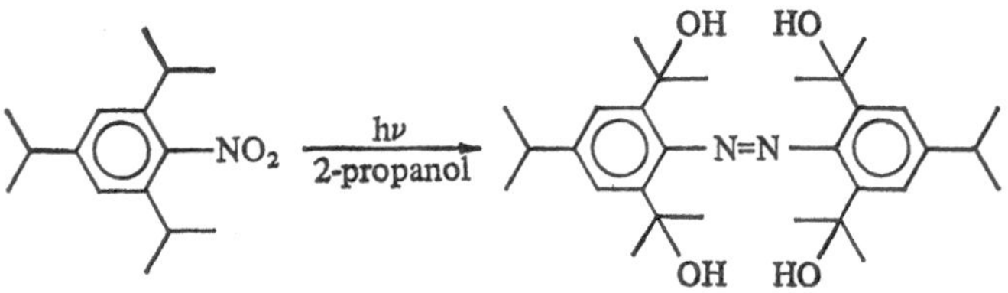

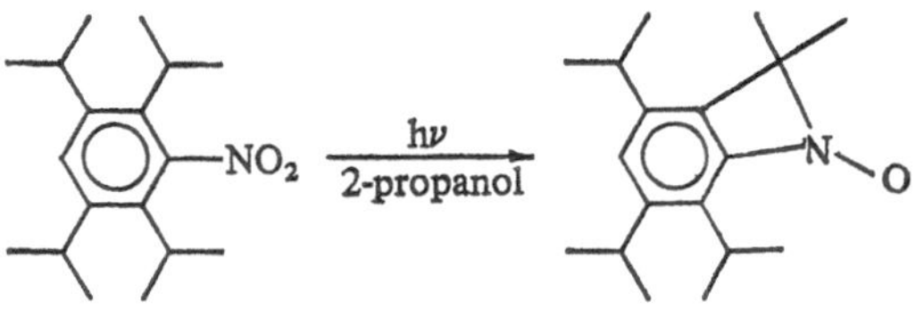

In view of the short lifetime ($\approx 10^{-9}$ s), which has been reported for nitro-benzene [13] and might be expected for similar nitroaromatics with low lying (n, π^*) states, and since intramolecular hydrogen abstractions from benzylic positions take place, lack of oxygen quenching may be inconclusive.

A few cases of intramolecular hydrogen abstractions from β-positions of the *ortho*–attached side chain have been published. 2-(2-nitrophenyl)-ethanol yields 1-hydroxy-3H-2-indolinone [51]:

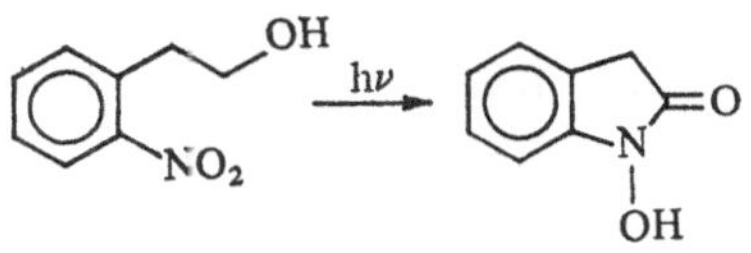

An intramolecular hydrogen abstractions has also been claimed [42] to occur from one methyl group of 1,3,5-tri-ethyl-2-nitrobenzene. The photochemistry of o-nitro-*tert*-butylbenzenes has been studied in some detail [42,48,52–60,62].

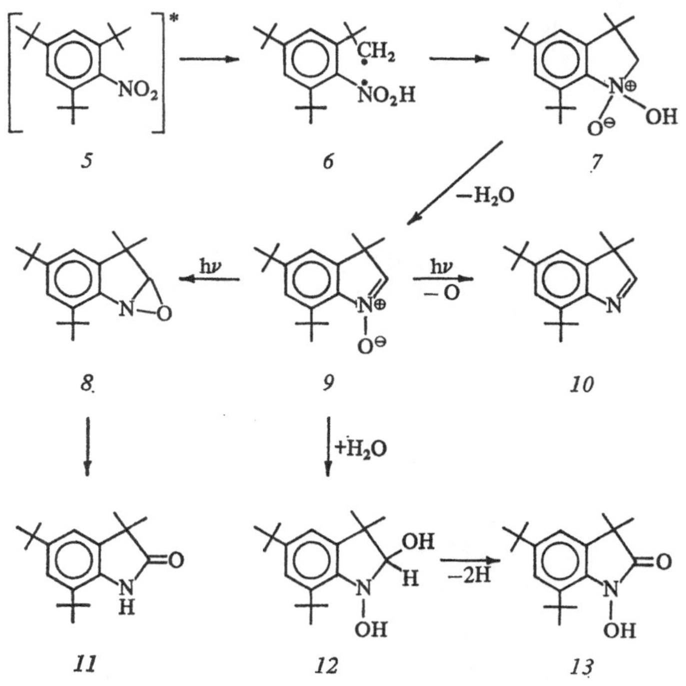

1,3,5-tri-*tert*-butyl-2-nitrobenzene (*5*) has been transformed into the cyclic hydroxamic acid *13* and the imine *10* upon irradiation ($\geqslant 280$ nm) in 2-propanol, no photoreduction by the solvent could be observed [42]. Other authors [52] reported the formation of the lactam *11* by photolysis (254 nm) of *5* in cyclohexane or benzene. They proposed the intermediacy of *6, 7, 8*, and *9* in the course of the reaction. The first isolable product, however, is none of the previously isolated ones but the nitrone *9* [56a)] when the irradiation of *5* is carried out in various solvents other than aliphatic amines, and it may be conveniently prepared in large quantities by irradiation of crystalline *5* [56b)]. *9* has been converted to *11* and other products as described elsewhere [61)], and it is oxygenated by a hydration/dehydrogenation sequence [56a)].

A variety of other 2-nitro-*tert*-butylbenzenes (*14*) has been transformed into N-hydroxy-2-indolinones (*20*) [48,53,58)]. In some instances, both cyclization modes of the intermediate diradical (*i.e.* modes a and b) are effective as derived from by-products. Nitrones (*18*) have not been isolated except for $R = (CH_3)_3C$ [55)].

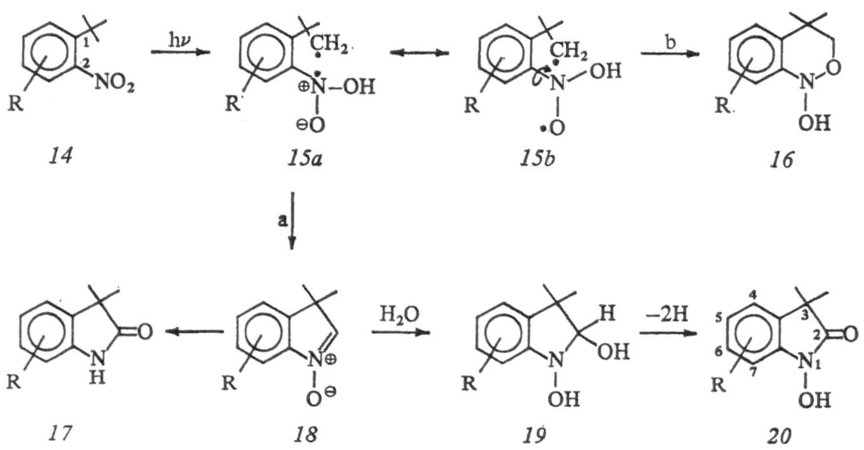

Intramolecular hydrogen abstraction leading to the diradical *15* has been shown to be rate-limiting in the sequence *14 → 15 → 18* by means of deuterium isotope effect studies. When the choice is given, hydrogen is abstracted 4.1—4.8 times faster out of a CH_3-group than deuterium out of a CD_3-group in the labeled compounds *21* and *22* [57)].

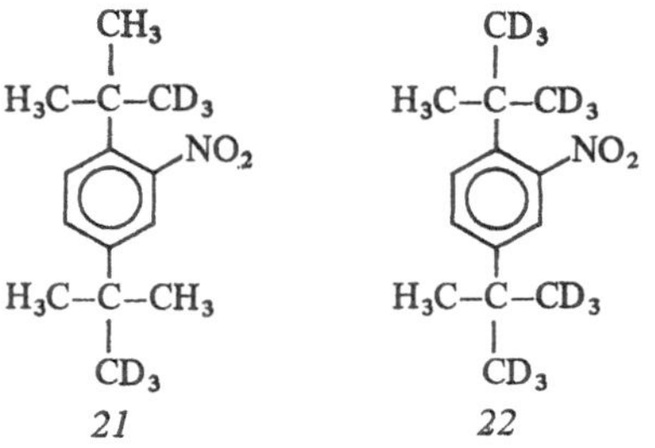

Sensitization by acetophenone [60] and benzophenone [48] could be demonstrated for the cyclization of *14*, R = 4—C(CH$_3$)$_3$ to *20*, R = 6—C(CH$_3$)$_3$. With the latter sensitizer, enhanced yields of N-hydroxy-2-indolinone formation have been observed relative to direct irradiation runs. Triplet benzophenone does not act as dehydrogenating agent in the step *19 → 20*, since the pinacolization of benzophenone by benzhydrol [31] could be quenched upon addition of *14* (R = 4—C(CH$_3$)$_3$).

Quenching studies on the photolysis of *14* (R = C(CH$_3$)$_3$) using 1,3-pentadiene reveal that even at 2.5 mole l^{-1} pentadiene not all excited states are quenched [59], thus at present a triplet state as reacting excited state cannot be regarded as proven for the direct irradiations although it seems likely in view of the high triplet yield [13] reported for nitrobenzene.

The excited state configuration is likely to be n, π* as judged from the capability of the excited nitro group to abstract hydrogens which are in no way activated. Lack of activation, however, is overcompensated by the close proximity of the potential reaction partners. Although quantum yields of disappearance of starting material would be highly desirable, so far only quantum yields of formation of products (*i.e.* N-hydroxy-2-indolinones) have been obtained.

The quantum yield for formation of *20*, R = 6-C(CH$_3$)$_3$, from *14*, R = 4-C(CH$_3$)$_3$, *via* photolysis in *tert*-butyl alcohol and oxidative workup, has been determined as $(1.2 \pm 0.1) \times 10^{-2}$ [48]. Quantum yields for formation of various other N-hydroxy-2-indolinones from appropriately substituted 2-nitro-*tert*-butylbenzenes fall into the range 1.1×10^{-2} to 2.2×10^{-2} [59] and thus are of the same order of magnitude as the disappearance quantum yield reported for nitrobenzene in 2-propanol [11].

No reaction at all is observed upon irradiation of *14*, R = 4-NH$_2$ [48] in methanol or *14*, R = 4- (or 5-) -OH in aqueous alkaline solution [59]. From solvent shifts in the absorption spectra and lack of reactivity a lowest (π, π*)- or charge transfer excited state is implied.

In aliphatic amines (diethylamine or triethylamine) the intramolecular hydrogen abstraction is quenched almost completely. Instead, smooth photoreduction of the nitro group without participation of the side chain is observed with 1,3,5-tri-*tert*-butyl-2-nitrobenzene (*5*) [56a] and *14*, R = C(CH$_3$)$_3$ [62]. Products derived from the respective phenylhydroxylamines were isolated in both cases. Again, an electron transfer, which does not seem to suffer from steric restrictions, is operative (see also Section A.I.3).

5. Addition of Nitrobenzenes to π-Systems

Addition reactions of excited nitro groups to alkenes and alkines, both inter- and intramolecularly, have been reviewed previously [8].

Following a report by Büchi and Ayer [63] on the photolysis of nitrobenzene in cyclohexene and 2-methyl-2-butene, a thorough study on nitrobenzene cycloaddition to olefines has been undertaken by de Mayo and co-workers [30,64]. Evidence was presented for the previously postulated [63] formation of 1,3,2-dioxazolidines:

D. Döpp

1,2,3	R
a	H
b	3–Cl
c	4–Cl

Whilst irradiation of nitrobenzene in cyclohexene at room temperature gives a complex mixture of products, low temperature irradiations at 100% conversion gave a crystalline solid (2a), which could by crystallized at $-80\,°C$ from ethyl acetate but rapidly decomposed at room temperature [64]. Catalytic hydrogenation yielded 90% *cis*-cyclohexane-1,2-diol (containing 5% *trans*-isomer). Spectral data also are in agreement with structure 2. From the decomposition products [63], the analogy of 2 with primary ozonides of olefins is not purely formal. The addition is to be classified as electrophilic attack on the olefin and may proceed through a triplet state, since sensitization with benzophenone (366 nm, $-15\,°C$) resulted in the same consumption of 1b as in unsensitized runs. In view of the high triplet yield of nitrobenzene [13], a triplet state reaction could be envisaged also for the direct irradiation. A stepwise mode of addition of 1a has been demonstrated by addition to *cis*- and *trans*-2-butene: in both cases after hydrogenation the same mixture of 65% *d,l*- and 35% *meso*-2,3-butanediol was obtained.

Whereas a straight Stern-Volmer plot could not be obtained in quenching studies with octafluornaphthalene, the orders of magnitude of the rate constants of triplet decay (k_d) and of reaction with olefin (k_r) could still be estimated: $k_d = 6 \times 10^8\ s^{-1}$ [a] and $k_r = 2 \times 10^7\ l\ mole^{-1}s^{-1}$ [b]. In view of this rapid decay, a high concentration of olefin is evidently necessary for effective addition.

Increase in electron availability (as measured by the ionisation potential) within the target olefin does indeed increase the rate of addition. Electron withdrawing groups (m-CN, m-Cl) in the nitrobenzene moiety stabilized the adducts, whereas an increased rate of decomposition was observed with adducts from p-chlorobenzene and m- or p-nitrotoluene [30].

1-Nitronaphthalene and 2-nitrobiphenyl did not undergo similar cycloadditions, and lack of reactivity was attributed [30] to the (π,π^*)-character of the lowest triplet state of these nitro aromatics.

On the other hand, aromatic nitro compounds having either lowest (n,π^*)- or (π,π^*)-triplets are able to induce oxidative ring cleavage of aromatic methoxy compounds [65–67]. A typical example is given below.

---

[a] $k_d = 10^9\ s^{-1}$ has been determined from more accurate measurements [13].

[b] This rate constant is twenty five times higher than the rate constant of hydrogen abstraction from 2-propanol: $k_H = 0.8 \times 10^6\ l\ mole^{-1}\ s^{-1}$ [14].

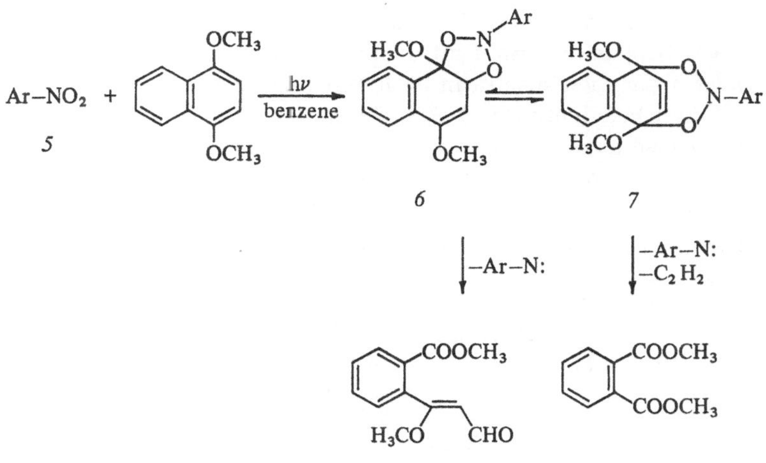

5 *6* *7*

8 70% (58%) *9* (20%)

Ar–N: ⟶ Ar–N=N–Ar + $Ar{-}NH_2$ Ar–NH–CHO

10 13% (–) *11* 5% (14%) *12* 7% (43%)

In addition to products *8*–*11*, varying amounts of N-formylanilines (*12*) are formed when methoxynaphthalenes are oxidized. Yields for Ar = 3-Cl-C_6H_4 are given along with those for Ar = 1-naphthyl (in parentheses).

The reaction is thought to proceed *via* an excited charge transfer complex to a thermal and prototropic equilibrium mixture of cycloadducts *6* and *7*, which decompose to the products shown.

The reaction depicted above proceeds also in other aprotic but not in protic solvents, and (π,π^*)-nitroaromatics (1-nitro-, 2-nitro- and 2-chloro-4-nitronaphthalene, 4-nitrobiphenyl) seem to be more reactive than several nitrobenzenes. However, the lifetime of the excited state of the respective nitroaromatic may play a dominating role in these cases.

II. Photoreduction of Nitronaphtalenes and Nitrobiphenyls

1. Spectroscopy of the Nitronaphthalenes

Phosphorescence data of several nitronaphthalenes, mainly from the recent literature [2–6, 68–70] have been compiled in Table 1. A thorough analysis of the absorption spectrum of 1-nitronaphthalene has also appeared [71].

In all cases, phosphorescence lifetimes (Table 1) are longer than a few milliseconds, which is a commonly accepted characteristic of (π,π^*)-triplets and markedly contrasts the short lived [13] and inefficient ($\phi_p < 10^{-3}$ [13]) phosphorescence of nitrobenzene[c].

[c] $\tilde{\nu}_{p\,(0-0)} = 21{,}100\ cm^{-1}$ and a lifetime of 0.1—1 s at 77 K in EPA have recently been reported [2] for nitrobenzene. See also Ref. [30].

Intrinsic lifetimes of at most 10 µs at room temperature have been estimated for the lowest triplets of both 1-nitronaphthalene and 2-nitronaphthalene from triplet–triplet absorption spectra in flash experiments [6].

At 77 K in EPA, lifetimes of 60 ± 10 ms for 1-nitro- and 242 ± 10 ms for 2-nitronaphthalene have been obtained from analysis of the transient absorption decay curves [6]. Differences to phosphorescence lifetimes under the same condi-

Table 1. Phosphorescence data (77 K) of five nitronaphthalenes

Nitro-naphthalene	Solvent	ϕ_p[1]	τ_p (ms)	$\bar{\nu}_{0-0}$ (cm^{-1})	ϕ_T	k_p (s^{-1})	k_d (s^{-1})	Ref.
1–	EPA	0.14	47	19,300	0.63 ± 0.1	4.6	17	[3]
	EPA			19,250				[2]
	Ethanol		50	520 nm				[68]
			50[2], 50 ± 2[4]	18,800[3]				
2–	EPA	0.26	206	19,900	0.83	1.5	3.4	[3]
	Ethanol		200	500 nm				[68]
			240 ± 6[4]	19,550[3] 492 nm[5]				
1,3–di	EPA	0.14	85	19,300	0.7[6]	2.3	9.5	[3]
				19,230[3]				
1,5–di	EPA	0.29	58	19,210	0.7[6]	7.2	10	[3]
			110[2]	19,125[3]				
1,8–di	EPA	0.79	174	20,000	0.8[6]	5.7	0.1	[3]

[1] with reference to benzophenone: $\phi_p = 1.00$.
[2] Ref. [69].
[3] Ref. [70].
[4] Ref. [6].
[5] Phosphorescence polarisation along long axis, in plane. Ref. [4].
[6] Approximate values, based on triplet counting experiments, see Ref. [3].

tions are attributed to the different experimental methods used [6]. The rate constant for endothermic triplet energy–transfer from 1-nitronaphthalene ($E_T = 55$ kcal mole^{-1}) to cis piperylene ($E_T = 57$ kcal mole^{-1}) has been reported: $k_{et} = 6.8 \times 10^7$ l moles^{-1} s^{-1} [13].

It is interesting to compare the data given for 1-nitro- and 2-nitronaphthalene. The nitro group in the former is twisted by 49° out of the ring plane as determined from an X-ray study [72]. Phosphorescence lifetime is one order of magnitude smaller, the triplet decay rate is five times greater and the long wavelength absorption (λ_{max} 330 nm) is shifted to a lower wavelength than in 2-nitronaphthalene (λ_{max} 350 nm).

2. Photoreduction of 1-Nitronaphthalene

Upon irradiation (366 nm) in 2-propanol, 1-nitronaphthalene is inefficiently photoreduced ($k_H < 10^2$ l moles^{-1} s^{-1}) despite its high triplet yield [13]. With tributylstannane, a marked increase of the hydrogen abstraction rate ($< 3 \times 10^6$

l moles^{-1} s^{-1}) was observed. These results chemically confirm the (π,π^*)-assignment of the lowest triplet level.

Reluctance to photoreduction shown by 1-naphthaldehyde has also been attributed to lack of reactivity of the lowest (π,π^*)-triplet state [73].

From photoreduction (> 280 nm) in diethylamine, low yields of 1-naphthylamine and the corresponding azo- and azoxy compounds have been obtained [16].

Photolysis (366 nm) in acidified 50% aqueous 2-propanol at varied HCl-concentrations results in remarkable enhancement of photoreduction compared to neutral 2-propanol. The highest disappearance quantum yield measured was 1.28×10^{-2} for 6 M HCl [74]. 4-chloro-1-naphthylamine is formed as main product [74,75].

3. Photoreduction of 2-Nitronaphthalene

In 2-propanol solution with 313 nm excitation, a quantum yield of 0.037 has been estimated for starting material disappearance [76]. 2-Naphthylhydroxylamine has been regarded as product in analogy to nitrobenzene [11] and spectroscopic arguments. Quenching experiments using 1,3-cyclohexadiene ($E_T = 54$ kcal mole^{-1}) show a fairly linear Stern-Volmer plot in the range of $0-13 \times 10^{-4}$ M diene concentration with 1.6×10^{-4} M 2-nitronaphthalene. No product formation was observed in aerated solutions. Since no T–T absorption for 2-nitronaphthalene could be observed in flash photolysis experiments, a triplet lifetime of 20 μs at room temperature is envisaged by these authors [76] (see Section II.1). From their data and ϕ_T from Table 1, the following limiting values for rate constants are derived for triplet 2-nitronaphthalene:

a) radiationless decay: 5×10^4 s$^{-1} < k_{dt} < 2 \times 10^6$ s^{-1}

b) hydrogen abstraction from 2-propanol:
$$2 \times 10^3 \text{ s}^{-1} < k_H \cdot [\text{2-PrOH}] < 8 \times 10^4 \text{ s}^{-1}$$

c) quenching: 6×10^7 l mole^{-1} s$^{-1} < k_q < 3 \times 10^9$ l mole^{-1} s^{-1}.

Although a slower hydrogen abstraction rate is anticipated for $^3(\pi,\pi^*)$-states compared to $^3(n,\pi^*)$-states, 2-nitronaphthalene is more efficiently ($\phi = 0.037$) photoreduced than nitrobenzene ($\phi = (1.14 \pm 0.08) \times 10^{-2}$ [11]). The longer triplet lifetime obviously overcompensates a slower reaction rate.

4. Photoreduction of Nitrobiphenyls

The triplet state of 4-nitrobiphenyl has been observed in laser flashed benzene solution (λ_{max} 540 nm, τ 10 ns at room temperature) [132]. 4-Nitrobiphenyl and 4,4'-dinitrobiphenyl have been photoreduced by sodium formate in buffered aqueous methanolic solution [43b], 15% 4-aminobiphenyl and 11% 4,4'-azobiphenyl as well as 49% 4-amino-4'-nitrobiphenyl and 20% 4,4'-(p-nitrophenyl)-azoxybenzene, respectively, could be isolated and identified by comparison with authentic samples.

The occurrence of at least one bimolecular reaction (besides quenching) of excited 4,4'-dinitrobiphenyl with methoxide ion has been established from linear

ϕ^{-1} vs. $[CH_3O]^{-1}$ plots and attributed to photoreduction of the nitro function-
ality [43 b]. From the translation of the original paper it does not become clear,
however, whether and how corrections have been made for photosubstitution
reactions, which have been shown to occur when nucleophiles react with excited
nitrobiphenyls (see Section B.II.2.).

III. Photoreduction of Nitropyridines and Nitropyridine-N-oxides

1. 4-Nitropyridine

The neutral molecule 4-nitropyridine (1) and the corresponding pyridinium ion
2 should exhibit different photochemical reactivity. 2 is expected to resemble
nitrobenzene since the nonbonding electrons on the ring nitrogen are not readily
available for excitation.

In 2-propanol solutions, acidified with hydrogen chloride, 1 is photoreduced
quantitatively to 4-hydroxylaminopyridine (3) as determined by following the
spectral changes during irradiation and workup under alkaline or neutral conditions to yield 4,4'-azo- or 4,4'-azoxypyridine [77].

Disappearance quantum yields for 1 were found to depend on the HCl-concentration [77,78]. In view of the analytical methods used, the values obtained by
Cu and Testa [78] with 313 nm excitation seem to be the more reliable ones (Table
2). It is noteworthy that quantum yields obtained in air-saturated solutions
(ϕ_{air}) seem to be less dependent on HCl than those (ϕ_{deg}) in degassed solutions;
which suggests that there might exist two different pathways for hydrogen abstraction.

Table 2. Quantum yields for disappearance of 4-nitropyridine
(1, 7.79×10^{-3} M in 2-propanol) with varied HCl-concentration

[HCl] M	ϕ_{deg}	ϕ_{air}	ϕ_{deg}/ϕ_{air}
0.1	0.27 ± 0.03	0.084 ± 0.02	3.2
0.3	0.31 ± 0.03	0.11 ± 0.01	2.8
0.5	0.70 ± 0.08	0.14 ± 0.02	5.8
1.0	0.56 ± 0.03	0.087 ± 0.02	6.4
2.0	0.59 ± 0.02	0.10 ± 0.01	5.9
3.0	0.16 ± 0.02	0.084 ± 0.01	1.9
4.0	0.082 ± 0.01	0.081	1.0
6.0	0.055 ± 0.007	0.063 ± 0.005	0.9

The highest ϕ_{deg}/ϕ_{air} ratio and the highest photoreduction yield are observed at roughly the same acid concentration. This suggests that the nitropyridinium triplet is very sensitive to oxygen [78].

On a preparative scale, the effect of oxygen (1 atm) could not be elucidated, since the reaction obviously took a different course [77]. 1,3-pentadiene (2.5×10^{-2} M) was found to retard the photolysis of *1* (2×10^{-3} M in 2-propanol acidified with HCl) [77].

No phosphorescence could be observed (thus $\phi_p < 10^{-3}$) from *1* in acidified 2-propanol or ethylene glycol/water (1:2) solutions upon 313 nm exitation at 77 K [78]. This is surprising in view of the oxygen effect cited above.

The proposed [15] electron transfer mechanism for this photoreduction parallels that given earlier for nitrobenzene and is supported by the observation of two first order transient absorptions ($\tau \sim 1$ ms) in flashed acidified 2-propanol solutions of *1*. These absorptions are assigned to the radical *4* and its conjugate acid *5*.

$$ H-\overset{\oplus}{N}\langle \bigcirc \rangle -N\overset{\bullet}{O}_2^{\ominus} \ +H^{\oplus} \ \rightleftharpoons \ H-\overset{\oplus}{N}\langle \bigcirc \rangle -N\overset{\bullet}{O}_2H $$

4 (480 nm)	*5* (430 nm)

In contrast to nitrobenzene, as significant inefficiency of photoreduction is observed [15] at high (6 moles l^{-1}) HCl concentrations, which has not been explained yet.

2. 4-Nitropyridine-N-oxides

It had been shown previously [79] that 4-nitropyridine-N-oxides *1 a—d* are photo-reduced (> 300 nm) in ethanol solution to the corresponding 4-hydroxylamino-pyridine-N-oxides *2 a—d*. Presence of oxygen was found to alter the course of the reaction.

1 e is not photoreduced, instead it is transformed into 4-hydroxy-3,5-dimethyl-pyridine-N-oxide *3* [80] very likely by a nitro-nitrite-isomerisation [8,81] pathway, which is favored since two flanking methyl groups enforce twisting of the nitro group out of the plane of the ring system. Since no oxygen quenching is observed in this reaction, it is proposed that an excited singlet state is responsible for this rearrangement.

Other authors [82,83] also found a delicate concentration and viscosity-dependent balance between a nitro-nitrite-isomerisation [8,81] route and photo-reduction governing the photochemistry of the unsubstituted pyridine-N-oxide (*1 a*).

The phosphorescence of a 5×10^{-2} M solution of biacetyl in de–aerated 2-propanol at room temperature could be quenched completely by *1 a, d, e* (10^{-3} M) [84]. In all three cases, the corresponding photoreduction products *2 a, d, e* emerge from analogous preparative scale biacetyl sensitized runs. Since *2 e* is also formed, steric hindrance to hydrogen abstraction from solvent cannot be too effective when a (probably longer-lived) triplet is populated, whereas it might be effective in the direct photolysis of *1 e* [80], where isomerisation competes with reduction probably in the (short-lived) singlet state.

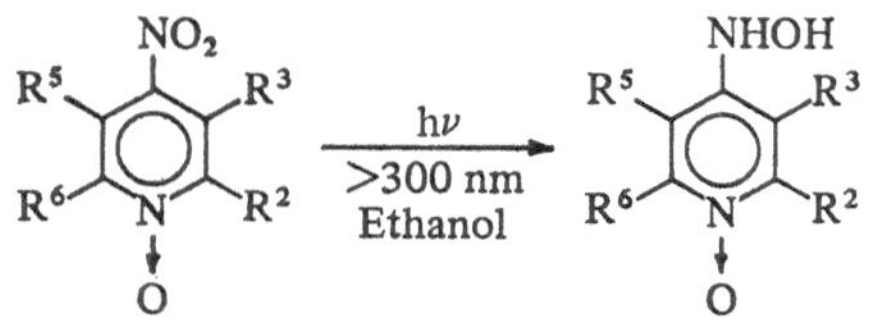

1a–e *2a–d*

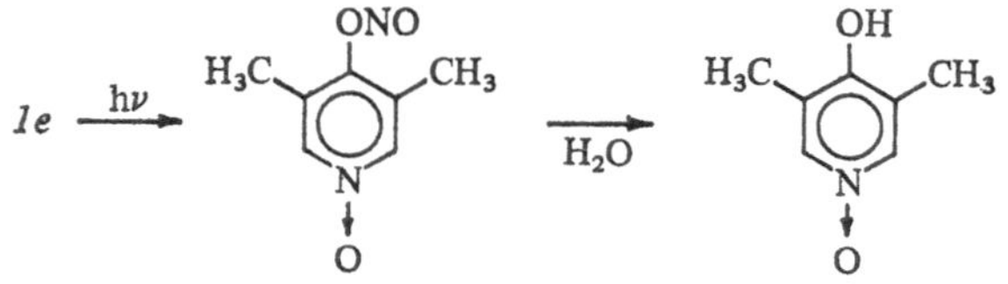

3

1,2	R^2	R^3	R^5	R^6
a	H	H	H	H
b	CH_3	H	H	H
c	CH_3	H	H	CH_3
d	H	CH_3	H	H
e	H	CH_3	CH_3	H

A (π,π^*)-configuration has been assigned to the lowest excited triplet state (E_T 52 kcal mole^{-1}) of 4-nitropyridine-1-oxide from both phosphorescence and S—T absorption studies [85].

B. Nucleophilic Aromatic Photosubstitutions

I. General Remarks

Nucleophilic aromatic photosubstitution in general is a rapidly growing field [86–89] which originated from research on light-induced reactions of aromatic nitro compounds but is by no means restricted to this class of compounds.

Most of the reactions [90–123] we shall deal with in this chapter may be generalized either as

$$O_2N{-}Ar{-}X^* + Y \longrightarrow O_2N{-}Ar{-}Y + X$$

with Y being a neutral or anionic nucleophile and X a leaving group or (strictly formally) hydrogen, or as

$$O_2N{-}Ar^* + Y^{\ominus} \longrightarrow Ar{-}Y + NO_2^{\ominus}.$$

In these bimolecular reactions the lifetime of the reacting excited state is a very important factor. In view of lifetime considerations, longer-lived triplets might be more likely candidates for the reacting excited states than singlets. Most nucleophilic aromatic photosubstitutions seem to proceed *via* (π,π^*)-triplet states, but cases of the intermediacy of singlet states are also known.

In most of the cases studied bimolecular kinetics are followed, the rate constants of product formation depending on the rate of light absorption and on nucleophile concentration. Triplet lifetimes (as determined from quenching studies) also depend on nucleophile concentration. This means that the excited state is quenched by the nucleophile, accompanied by either product formation or reversal to starting material. In view of the inherently different triplet lifetimes of different substrates, it is highly desirable to rely on rate constants rather than on quantum yields in studying the reactivity of excited nitroaromatics towards nucleophilic photosubstitution.

Aromatic nitro compounds, which have been subject to nucleophilic photo-substitution, are

a) nitrobenzene, dinitro- and halonitrobenzenes [95,97–99,105],

b) nitrophenylphosphates [90,113,114] and sulfates [90],

c) various nitrophenyl ethers [91–96,99–101,103,107,109,114–121],

d) 1-nitronaphthalene [104,116,120,121] and 1-nitroazulene [104],

e) nitrobiphenyls and nitroflucrenes [106],

f) halonitro- [110] and methoxynitronaphthalenes [68,102,110,116,120–122],

g) Nitroheterocycles [111,123].

II. Orientation Rules and Representative Examples

Substituents on an aromatic ring may show activating and directing effects in electrophilic substitution. The same situation is encountered in nucleophilic photosubstitution.

Four general trends [86,87] are observed and will be illustrated with a few typical examples.

a) m-Activation by a Nitro Group: In striking contrast to ground state chemistry, various leaving groups (or hydrogen) in m-position relative to the nitro group are replaced by nucleophiles, as is demonstrated with 3-nitroanisole:

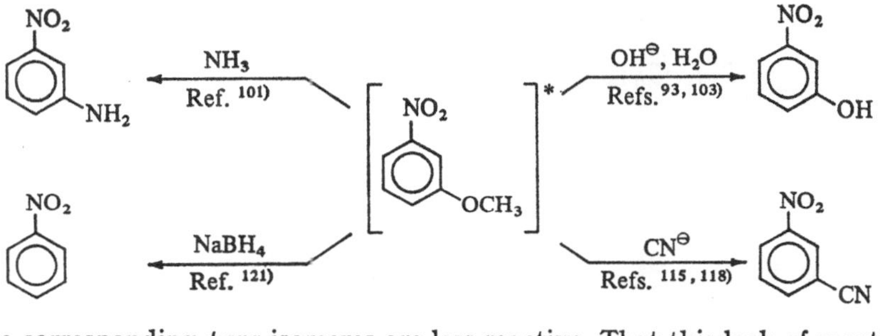

The corresponding *para*-isomeres are less reactive. That this lack of reactivity is not due to a shorter excited state lifetime, is demonstrated with 4-nitrovera-trole, in which specifically the m-methoxy group is replaced upon irradiation in the presence of the respective nucleophiles:

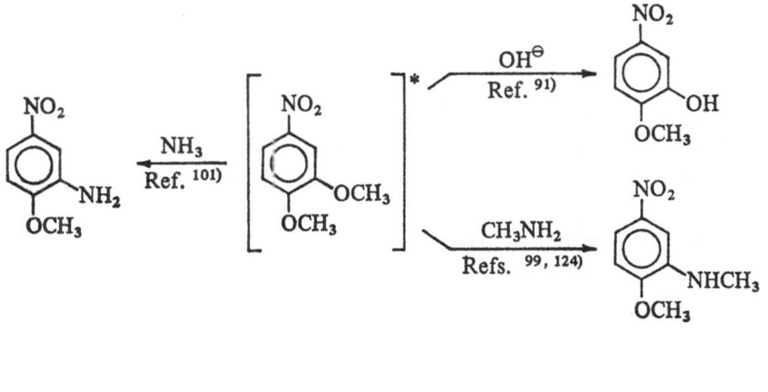

The *"meta*-activation-rule" also holds for naphthalene derivatives [68,102,122] and among these also for those cases where nitro group and leaving group (mostly methoxy) are attached to different rings but in positions (*e.g.* 1,6-, 2,4-, 2,5-, 2,7-) which are equivalent to *"meta"*. Thus 1-methoxy-3-nitro- and 2-methoxy-7-nitronaphthalene are photohydrolyzed by aqueous alkali to the corresponding nitronaphthols whereas 1-methoxy-5-nitro-, 2-methoxy-6-nitro- and 1-methoxy-7-nitronaphthalene are not [68,102].

b) α-Reactivity: Pronounced preference for nucleophilic attack at the α-position in bicyclic and tricyclic aromatics (*i.e.* position 1 in naphthalene or azulene, position 4 in biphenyl) is frequently observed. For example, introduction of a nitrile group into these positions seems to be typical for the hydrocarbons mentioned [125,127], and nitro groups in these positions are easily replaced by nucleophiles:

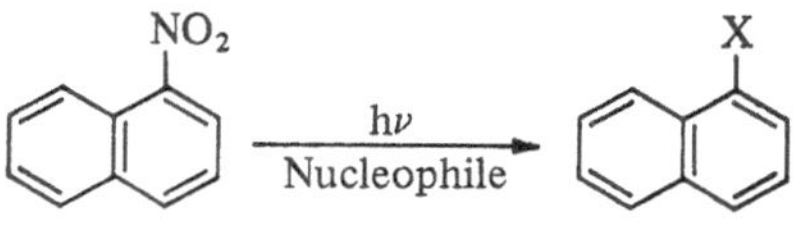

Nucleophile	$-X$	Ref.
$CH_3O^\ominus$	$-OCH_3$	104)
$CN^\ominus$	$-CN$	116,120)
pyridine	$-\overset{\oplus}{N}C_5H_5$	120)
$BH_4^\ominus$	$-H$	121)

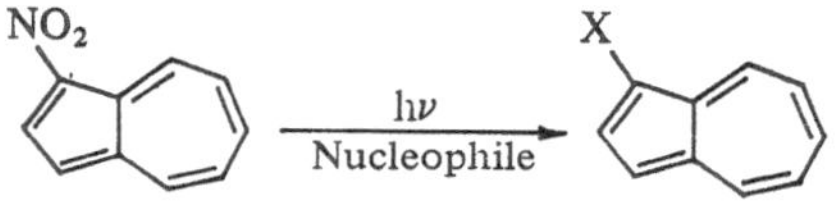

Nucleophile	$-X$	Ref.
$CH_3O^\ominus$	$-OCH_3$	104)
$CN^\ominus$	$-CN$	104)

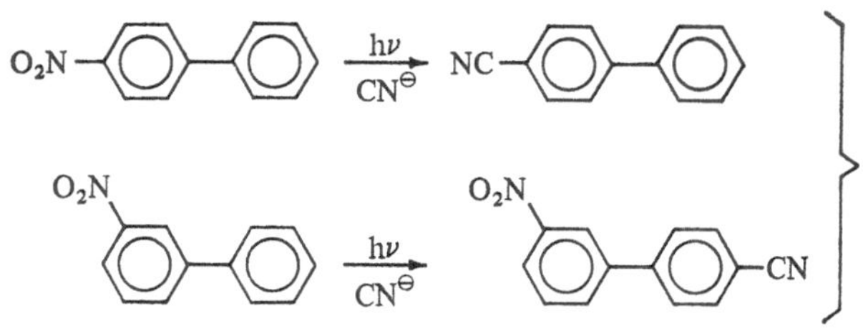

Refs. [106, 126]

c) Merging Resonance Stabilization During Product Formation: This trend is exemplified in those cases where an electron donor substituent is introduced photochemically into *ortho* or *para* positions relative to an electron acceptor substituent (here: nitro) [94,95,97,100,101] or *vice versa* [113,114,119]. The former case is illustrated in the photoamination [97,101] of nitro benzenes *1 a—f* to nitroanilines *2 a—d,f* and *3 a—e*, the latter in the nitrite displacement from p-nitrophenyl-esters and -ethers:

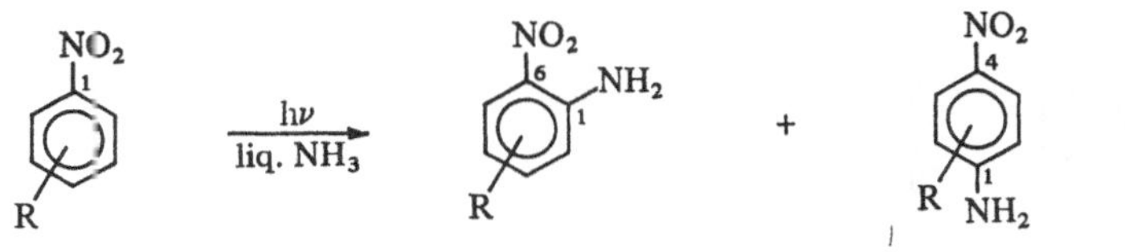

1	R
a	H
b	2-Cl
c	3-Cl
d	4-Cl
e	3-NO$_2$
f	2-NO$_2$

2	R
a	H
b	H (trace)
c	4-Cl (trace)
d	3-Cl (10%)
—	—
f	H

3	R
a	H
b	3-Cl
c	2-Cl
d	H (45%)
e	2-NO$_2$
—	—

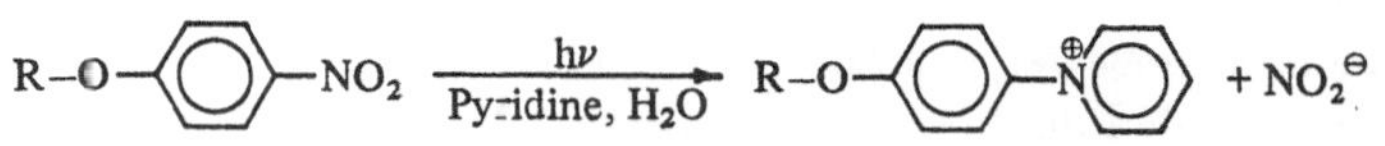

(R = —PO$_3$Na$_2$ [113]), —CH$_3$ [114]), various aryl groups [119,120]).

Tendency to achieve resonance stabilization also seems to be effective in the three different modes of photosubstitution encountered with 2-fluoro-1-methoxy-4-nitronaphthalene [110]:

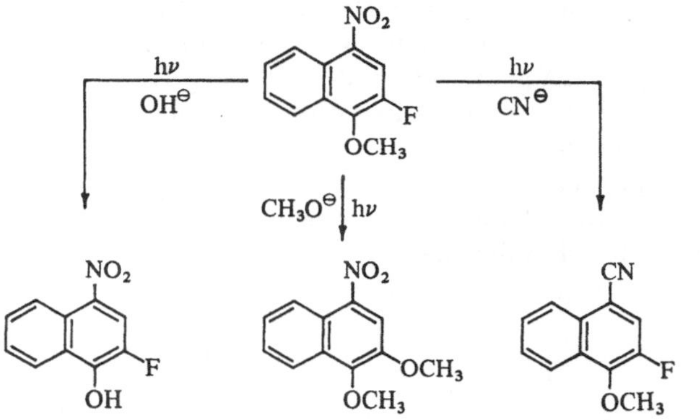

α-Activation and directive effects of the nitro- and methoxy groups may play a role, too.

d) Ortho-/Para-Activation by Methoxy Groups: This effect is outside the nitro-aromatic series demonstrated in the photocyanation of anisole (53% o- and 47% p-methoxybenzonitrile emerge as products) [127]. Its interplay with the other effects mentioned seems to be documented in a number of cases of photosubstitutions of methoxynitroaromatics [103,105,106,110,116–118,120,121].

71

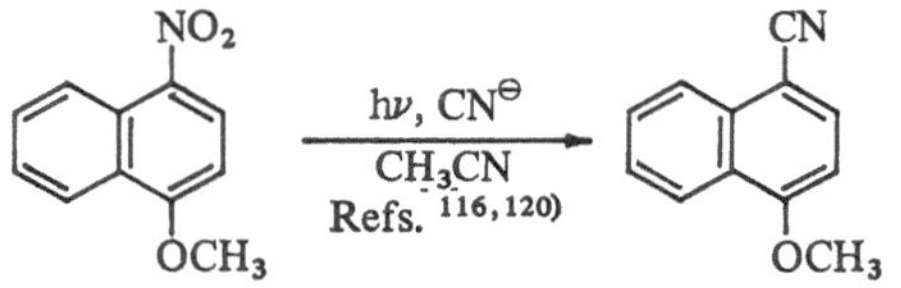

Substitution in α-position and *para* to methoxy.

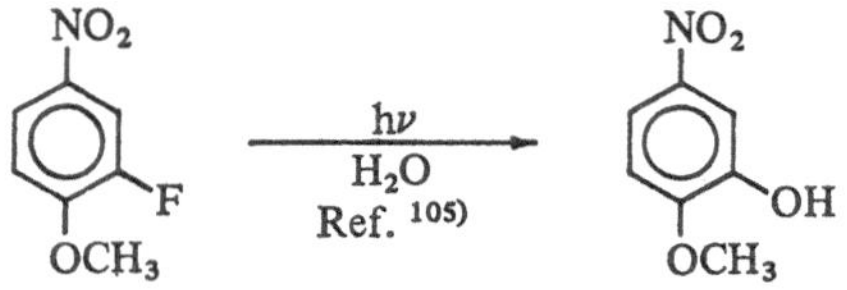

Combined activation by nitro and methoxy makes fluorine susceptible to substitution by water.

An additional *ortho*-directing effect by the methoxy group at C-1 certainly contributes to the specific reactivity of 4-nitroveratrole [91,99,101,124] [see above in Section a)].

Other factors, however, should also be effective. A specific influence of the solvent [101,116], added detergents [120] and remote electron donating substituents [119] has been observed. Steric hindrance, which certainly is of influence in the nucleophilic photosubstitution reactions of α-nitronaphthalenes, has been found to alter the reactivity of nitroanisoles [59].

III. Details of Nucleophilic Aromatic Photosubstitutions

1. Nitrobenzene Derivatives

The photohydrolysis of m-nitroanisole (mNA) has been thoroughly studied. The quantum yield of m-nitrophenol (mNP) formation reaches a limiting value of 0.23 at 0.07 M $OH^\ominus$, and there is no wavelength dependency at pH 12 over the 254—334 nm range [93]. A linear plot of ϕ_{mNP}^{-1} vs. $[^\ominus OH]^{-1}$ justifies the assumption of a bimolecular reaction of $OH^\ominus$ with the reacting excited state of mNA [93]. The kinetics are consistent with complex formation between (π,π^*) excited mNA, the complex in turn may either revert to starting material or continue to mNP. From ^{18}O tracer studies it follows unambiguously that the aryl-oxygen bond is broken in the product-forming step.

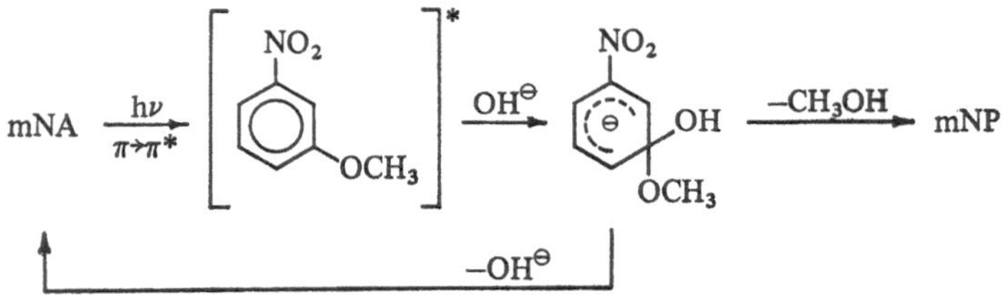

This reaction can be sensitized by benzophenone or benzophenone disulfonate with 254 nm irradiation [109]. Rate constants of 4.3×10^8 l mole^{-1} s^{-1} and 4.4×10^8 l mole^{-1} s^{-1} are obtained for the sensitized and unsensitized photohydrolysis under a nitrogen atmosphere. With admission of air, both the sensitized and direct photohydrolysis are diminished [109]. The reaction is efficiently quenched by piperylene ($> 4 \times 10^{-2}$ M). Using 3,3,4,4-tetramethyldiazetine dioxide [128] as quencher, a lifetime (in the absence of quencher, but in the presence of 10^{-2} M NaOH) of 24 ns is derived for the reacting $^3(\pi,\pi^*)$-state. Upon flashing with a frequency–doubled ruby laser, a species (λ_{max} 400 nm) with a lifetime of 40 ns in the presence of 4×10^{-2} M OH and 670 ns in the absence of OH$^\ominus$ has been detected [130].

Flash experiments [109] show completion of mNP formation within the flash duration time (20 μs). Therefore the short-lived species (λ_{max} 340—350 nm and 490—510 nm, lifetime $\sim$ 40 ms in presence of OH$^\ominus$ and 0,3 ms in presence of methylamine) observed cannot be an intermediate and is probably the radical anion of mNA [109].

In the analogous photohydrolysis of 3,5-dinitroanisole (dNA) at 313 nm a limiting quantum yield of formation of 3,5-dinitrophenole (dNP) of 0.48 has been obtained at 0.015 M NaOH in water containing 2% methanol. This remarkably clean reaction is sensitized by benzophenone [107] and thus may proceed *via* triplet excited dNA. From the absorption spectra of dNA in solvents of varied polarity [108] it may be concluded that close-lying (n,π^*)- and (π,π^*)-states exist within the singlet manifold. Since the singlet–triplet slitting of corresponding (π,π^*)-states is known to be larger than that of (n,π^*)-states, the lowest triplet of dNA is likely to have (π,π^*-)character [107].

A lifetime of 27 ns at room temperature (in the absence of quencher, but in the presence of 0.025 M OH$^\ominus$) has been calculated from a linear Stern-Volmer plot using 9-fluorenone as quencher [107]. In general, lifetimes of excited substrates are dependent on the nucleophile concentration. Quenching of the excited state by the nucleophile probably takes place by either formation of a σ-complex or simply return to ground state starting material.

Upon flashing (6 ns pulse) an alkaline solution of dNA in acetonitrile/water 1:1 with a frequency–doubled ruby laser (347 nm) three short-lived species have been detected [107]:

A) λ_{max} 550—570 nm, τ 40 ms
B) λ_{max} 412 nm, τ 500 ns
C) λ_{max} 475 nm, τ 12 ns.

During the decay of (A) no increase in dNP$^\ominus$ formation was noted, and dNP$^\ominus$ formation was found to be complete in shorter periods than 20 μs. Thus (A) is no intermediate, but has instead been identified, by comparison with ESR-results, as the radical anion of dNA.

Unfortunately, transient (B) absorbs almost exactly at the same wavelength as the final product dNP$^\ominus$ (λ_{max} 400—410 nm). It is very likely a precursor of the radical anion, and it is considered to be an aromate–nucleophile complex, which may decay to dNA, dNP or the radical anion (transient A). The latter ultimately may lead to photoreduction products, the major portion of it, however, returns to starting material.

The 475 nm species (C) is observed with even longer lifetime (55 ns) in the absence of OH$^\ominus$. Since the lifetimes observed are quite similar to that from a quenching study and since obviously this transient is quenched by OH$^\ominus$, this species is assigned to the triplet state of dNA. The sum of its rate constants for reaction with and quenching by $OH^\ominus$ was calculated to $8 \times 10^8 \, l \, moles^{-1} \, s^{-1}$ [107]. Thus complex formation is only one order of magnitude slower than a diffusion-controlled process.

The following scheme for the photohydrolysis of dNA has been suggested [108], [86]:

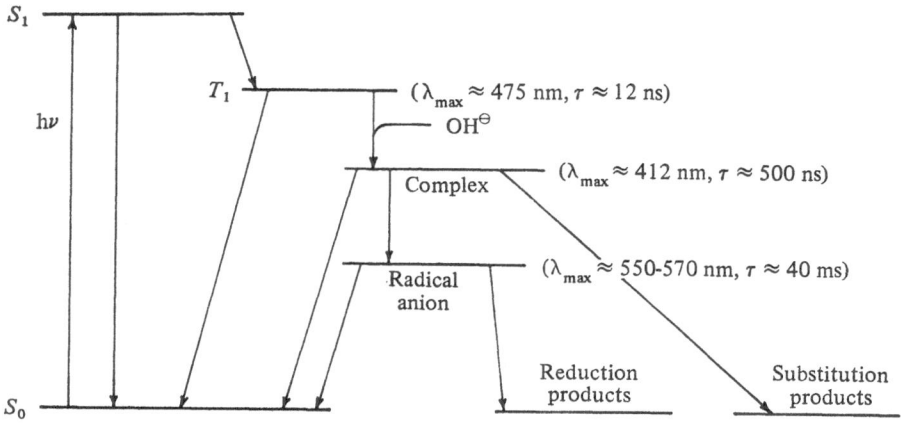

Schematic representation of the photohydrolysis of 3,5-dinitroanisole [86,130]

The photohydrolysis of 2-fluoro-4-nitroanisole to 2-methoxy-5-nitrophenole is sensitized by benzophenone and completely quenched by sodium sorbate [105]. The excited state multiplicity in photoaminations has also been studied. Photolysis of mNA in liquid ammonia yields m-nitroaniline. If the amination is carried out in a large excess of benzophenone, 2-methoxy-4-nitroaniline is formed instead and thus an excited singlet state as reacting species is envisaged in the unsensitized photoamination [100,101]. It may well be that uptake of the nucleophile present in high concentration successfully competes with intersystem crossing.

A competition between substitution and two kinds of quenching processes, either dependent or independent on the added nucleophile, is envisaged for the photosubstitutions of p-nitroanisole (pNA) [114,118]. With cyanide ions in aqueous aerated solution pNA is transformed on irradiation into 2-cyano-4-nitroanisole in high yield [115,118], probably *via* addition of cyanide to excited pNA.

$$pNA \xrightarrow{h\nu} [pNA]^* \xrightarrow{CN^\ominus}$$

Oxygen is needed to assist in removal of hydrogen from the complex [115,118].

The photohydrolysis of pNA ($E_T = 59.5$ kcal mole^{-1} [4]) is sensitized by benzophenone. The pseudo first order rate constant is found to be four times higher than in the unsensitized runs. The product pattern, however, was strictly the same in both the sensitized and unsensitized photohydrolysis, and this result strongly suggests that the triplet state is an intermediate also in the unsensitized reaction [117].

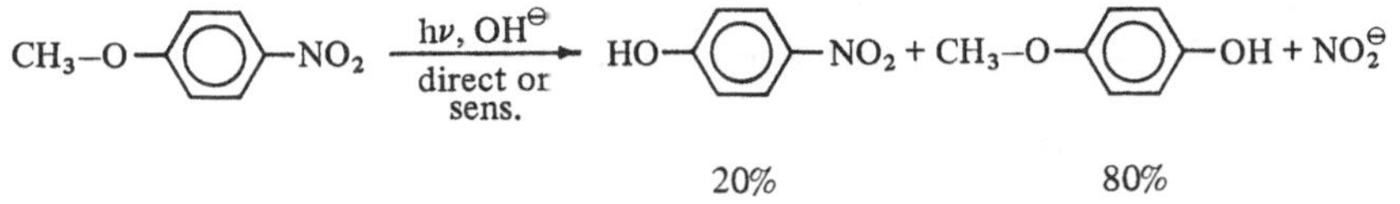

$$CH_3-O-\underset{}{\bigcirc}-NO_2 \xrightarrow[\substack{\text{direct or} \\ \text{sens.}}]{h\nu,\ OH^\ominus} HO-\underset{}{\bigcirc}-NO_2 + CH_3-O-\underset{}{\bigcirc}-OH + NO_2^\ominus$$

20%   80%

2. Nitronaphthalenes, Nitroazulenes and Nitrobiphenyls

Phosphorescence data and quantum yield of nitronaphthol formation have become available for various methoxynitronaphthalenes [68,108] (see Table 3).

Table 3. Phosphorescence data and quantum yields (ϕ) of nitronaphthol formation in photohydrolysis of some methoxynitronaphthalenes

Naphthalene deriv.	Type[1]	Phosphorescence (77 K) in Ethanol			Photosubstitution	
		0—0 nm	τ(ms) $\pm 10\%$	Ref.	ϕ[2]	Ref.
(α-Nitronaphthalenes):						
2-methoxy-5-nitro-	m	540	20	[68]	0.054[2]	[68,129]
1-chloro-2-methoxy-						
5-nitro-	m	530	20	[68]		
1-methoxy-4-nitro- [3]	p	520	50	[68]		
		520[4]		[122]	0.1[5]	[122]
1-methoxy-5-nitro-	p	510	50	[68]	0[2]	[68,129]
2,3-dimethoxy-5-nitro-					0.085[6]	[102]
(β-Nitronaphthalenes):						
1-methoxy-3-nitro-	m	545	15	[68]	0.109[2]	[68,129]
1-methoxy-6-nitro-	m	530	35	[68]	0.158[2]	[68,129]
2-methoxy-6-nitro-[3]	p	495	200	[68]	0[2]	[68,129]
2-methoxy-7-nitro-	m	495	200	[68]	0.026[2]	[68,129]
1-methoxy-7-nitro-	p	530	180	[68]		
2,3-dimethoxy-6-nitro-					0.102[6]	[102]

[1] See Section B.II.a for discussion of activation effects. m = *"meta"*, p = *"para"*.
[2] 313 nm, 1.4 N NaOH.
[3] Fluorescence in acetonitrile has been observed, see Ref.[122].
[4] In EPA.
[5] 365 nm, a mixture of 40% acetonitrile and 60% 0.1 N NaOH was used.
[6] 313 nm, dimethylsulfoxide/2 N NaOH.

Methoxynitronaphthalenes with *"para"*-orientation of the substituents indeed show a remarkable lack of reactivity compared to their *"meta"*-oriented isomers [68]. Within the pair of isomeric dimethoxynaphthalenes the preferred site of reactivity is always the *meta* position with respect to the nitro group: in 2,3-dimethoxy-6-nitronaphthalene, the methoxy group at position 3, in the 5-nitro-isomer, methoxyl at position 2 is replaced by hydroxyl [102]. This preference in the site of reactivity corresponds nicely to calculated[d] charge densities for the first excited singlet state, when charge densities for the ground and first and second excited singlet state are compared.

Since neither sodium sorbate nor fluorenone were effective as quenchers, the direct photohydrolysis of 2,3-dimethoxynaphthalene is likely to proceed *via* an excited singlet state. However, sensitization by benzophenone has been achieved [102].

Whereas the photohydrolysis of 1-methoxy-3-nitronaphthalene is quenched with tetramethyldiazetine dioxide (TMDD), and a triplet lifetime of 4 ns in the absence of quencher could be derived, no such quenching could be observed for 1-methoxy-6-nitronaphthalene [130]. Efficient sensitization of photohydrolysis of 1-methoxy-4-nitronaphthalene has been reported [122].

Phosphorescence and photohydrolysis of thirteen isomeric fluoronitronaphthalenes have been investigated [130]. Except for 1-fluoro-8-nitronaphthalene, fluorine substitution does not seem to alter the spectral characteristics of the parent nitronaphthalenes. Efficient substitution of fluorine by hydroxide or methoxide has been achieved in those cases where fluorine was positioned *"meta"* to nitro and/or was occupying an α-position [130]. Both functionalities may indeed be located in different rings without affecting the reactivity: The same quantum yield for photohydrolysis (0.33 at 366 nm, 0.1 N NaOH) and the same rate constant of deactivation of the excited state by reaction with hydroxide ($k = 1 \times 10^8$ 1 mole^{-1} s^{-1}) have been obtained for 1-fluoro-3-nitro- and 1-fluoro-6-nitronaphthalene, respectively. Photohydrolysis of the latter two compounds is quenched by TMDD, thus a triplet state is made likely as reacting excited state [130]. Efficient sensitization of 1-fluoro-3-nitronaphthalene photohydrolysis with benzophenone has been reported [110].

Quenching by TMDD has also been observed for the photomethoxylation of 1-nitroazulene [104], (see Section B.I) and the photocyanation of 4-nitrobiphenyl [126]. Thus these substitutions also probably proceed *via* excited triplet states.

The photocyanation of 1-nitroazulene was accompanied by a dark reaction and was thus less suitable for mechanistic studies than the methoxylation. 2-Acetonaphthone ($E_T = 59$ kcal mole^{-1}) and 2-naphthophenone ($E_T = 60$ kcal mole^{-1}) were found to be efficient sensitizers for the disappearance of 1-nitro azulene in the methoxylation reaction. The measured rate of product formation, however, seemed to be much lower than the rate of starting material disappearance. This effect has been attributed to sensitized decomposition of the product [104].

---

d) The calculations were carried out by J. J. C. Mulder, Leiden, according to the Pariser-Parr Pople method and were rounded off by taking into account configuration-interaction of all singly excited states. See Ref. [102] for results.

No reaction of excited 1-nitroazulene with hydroxide and cyanate ions or pyridine could be detected [104].

3. Nitrosubstituted Heterocycles

Some nucleophilic photosubstitutions of heterocycles have become known. 4-Nitropyridine-N-oxide, when irradiated in ethanol containing piperidine, is converted into 4-piperidino-pyridine-N-oxide [111] with an efficiency dependent on the concentration of piperidine. No indication of the multiplicity of the reacting excited state is given, however.

2-Nitropyrrole does not seem to undergo any photosubstitution with either methoxide, cyanate, cyanide or water [123]. 2-Nitrothiophene (*1 a*) and 5-bromo-2-nitrothiophene (*1 b*) undergo photocyanation smoothly and efficiently. The disappearance quantum yield for *1 a,b* equals the product quantum yield.

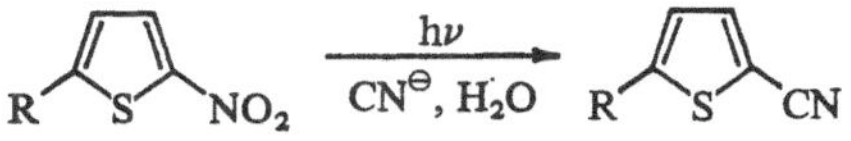

1, 2	ϕ_{313} [123]
a: R = H	0.39
b: R = Br	0.59

Again, a linear relationship of ϕ^{-1} and $[CN^{\ominus}]^{-1}$ shows a bimolecular reaction between the excited triplet state of *1* and the nucleophile to take place. The triplet lifetime of *1 a* is 4.7×10^{-7} s in water and 1.2×10^{-8} s in aqueous 10^{-2} M solutions of potassium cyanide as determined from quenching studies [123]. The nitro group in *1 a* is likewise replaced photochemically by methoxide and cyanate ions.

The case of 2-nitrofuran is especially interesting. The quantum yield of disappearance of starting material in the photocyanation reaction is 0.51 at 313 nm and not dependent on the cyanide ion concentration. The quantum yield of product formation, however, is dependent on the concentration of cyanide, a limiting value of 0.51 is reached at approximately 1 mole l^{-1} cyanide. Kinetics are in agreement with the formation of an intermediate X (the nature of which needs to be clarified) which is subsequently intercepted by a nucleophile. Water competes with cyanide in this product-forming step. This cyanation has been both sensitized and quenched, thus very likely it proceeds *via* a triplet state.

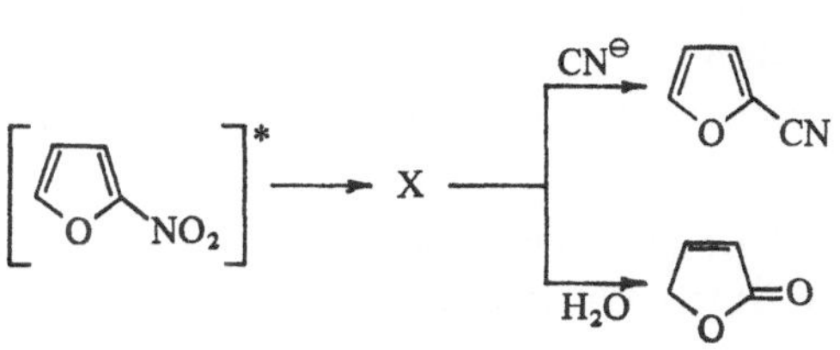

Replacement of a nitro group by methoxyl has also been reported to occur in N-butyl-5-nitro-2-furamide upon irradiation in methanol [131].

C. Other Light-Induced Reactions of Aromatic Nitro Compounds

1. Isomerizations

Nitro substituted stilbenes have been used as models in stilbene $cis \rightleftarrows trans$-isomerization studies. It had been reported previously, that $cis \rightarrow trans$-isomerization was effected within an exciplex of 4-nitrostilbene and triplet zinc etioporphyrin in benzene solution, and a photostationary state of 99.5% *trans* isomer had been reached [132].

Laser flash photolysis (30 ns, 50 mJ, 347 nm) of 4-nitrostilbene, 4,4'-dinitrostilbene, 4-nitro-4'-methoxystilbene and 4-dimethyl-amino-4'-nitrostilbene permits the observation of transients [133]. For the latter two compounds, the lifetime and the absorption spectra of the transients vary strongly with the polarity of the solvent used. First order decay rate constants are given in Table 4.

Table 4. First order rate constants ($k \times 10^{-6}$ s^{-1}, $\pm 10\%$) for the decay of transients from various nitrostilbenes in solution at room temperature [133]

Solvents used	Stilbenes subjected to Laser flash photolysis						
	4-NO$_2$		4,4'-di-NO$_2$		4-NO$_2$-4'-OCH$_3$		4-NO$_2$-4'-N(CH$_3$)$_2$
	cis	trans	cis	trans	cis	trans	trans
Cyclohexane	13	16	12.1	12.5	12.3	11.3	2.3
Benzene	12	15		9.8	8.0	7.2	0.48
Ethanol			9.6	9.2	3.6	3.5	0.25[1]
Methanol	12	12				3.0	
Dimethyl-formamide	8.9	9.6	5.1	4.8	1.75	1.75	0.047[1]

[1] By energy-transfer from triphenylene.

Identical spectra and lifetimes were observed for the first three compounds listed above. It is assumed, that the observed transients are triplet states of the nitrostilbenes on the basis of the following results:

1. Diffusion-controlled quenching of the transients is observed with oxygen, azulene and ferrocene,
2. the transient from 4-dimethylamino-4'-nitrostilbene is also formed by energy–transfer from triphenylene,
3. the lifetime of the transient from 4-methoxy-4'-nitrostilbene is identical with the lifetime of the expected triplet derived from the influence of azulene and ferrocene [134] on the position of the photostationary state [135] under the assumption of a triplet route for the $trans \rightarrow cis$- and a singlet route for the $cis \rightarrow trans$-isomerization.

It is concluded that there is no common intermediate in the direct $cis \rightarrow trans$- and $trans \rightarrow cis$-isomerizations of the nitrostilbenes.

Since similar transients are not observed from 4-cyano-4'methoxystilbene, 4,4'-dimethoxystilbene, 4-aminostilbene and 4,4'-dimethylaminostilbene, the possibility of observing such transients in the nitrostilbene series is certainly due to the tendency of the nitro group to enhance $S_1 \rightarrow T_1$ transitions [1].

It should be mentioned that only inefficient photocyclizations to the respective dihydrophenanthrenes have been observed with 4-nitrostilbene ($\phi = 0.0007$) and 3-nitrostilbene ($\phi < 0.0001$) compared to more efficient ($\phi = 0.28$) cyclization in 3-aminostilbene [136].

Non-coplanarity of the aromatic (or olefinic) system and the nitro group attached to it has been thought to be a prerequisite to an efficient nitro-nitrite rearrangement [81] (see ref. for a review of this well-known reaction). Exceptions to this rule have been found:

The photolysis of 4-nitroanisole in degassed acetonitrile or benzene yields 4-nitrosoanisole and 2-nitro-4-methoxyphenol [45]. Triphenylene ($E_T = 67$ kcal mole^{-1}; 4-nitroanisole: $E_T = 59.5$ kcal mole^{-1} [4]) has been used to sensitize the reaction, which is suppressed completely by nitric oxide. A rationale for the formation of the products observed is given below.

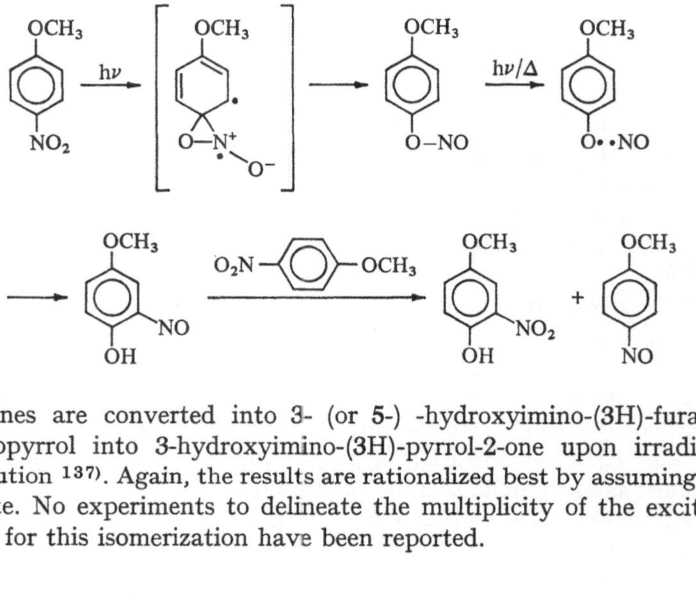

2-Nitrofuranes are converted into 3- (or 5-) -hydroxyimino-(3H)-furan-2-ones and 2-nitropyrrol into 3-hydroxyimino-(3H)-pyrrol-2-one upon irradiation in acetone solution [137]. Again, the results are rationalized best by assuming a nitrite intermediate. No experiments to delineate the multiplicity of the excited state responsible for this isomerization have been reported.

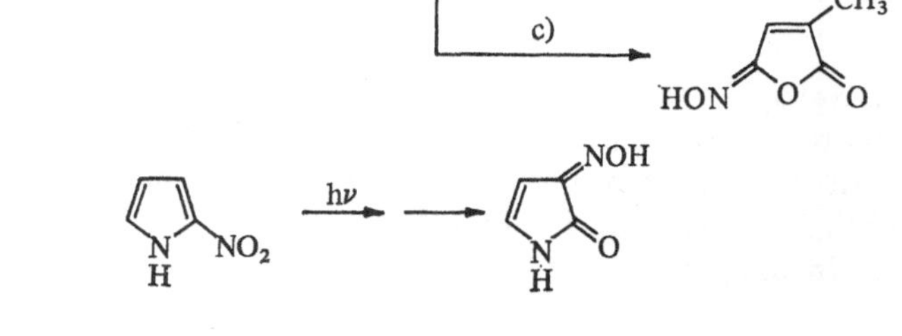

a) $R^1 = R^2 = H$
b) $R^1 = H$, $R^2 = CH_3$
c) $R^1 = CH_3$, $R^2 = H$

The photochemistry of 4-nitrobenzaldehyde [138] has been reinvestigated [139]. In aqueous solution, 4-nitrosobenzoic acid is formed. A ketene has been proposed as intermediate [139].

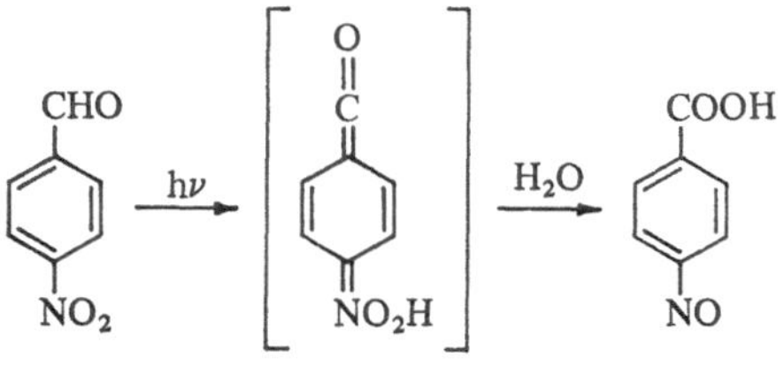

2. Photosubstitutions

Besides the well-established nucleophilic photosubstitutions of various groups in aromatic nitro compounds, a small number of other substitutions at excited aromatic nitro compounds have become apparent.

a) In deutero-trifluoroacetic acid in the dark, 4-nitroanisole undergoes H/D exchange only at the positions *ortho* to the methoxy group. Upon illumination, however, H/D exchange is accomplished at equal rate at all four unsubstituted positions in this molecule. Calculated charge distributions rather than localization energies in the lowest excited singlet state satisfactorily match the substitution pattern [140].

Under the same conditions nitrobenzene shows a *para > meta » ortho* preference in light-induced deuterium incorporation [140].

b) The case of unimolecular fission of 3-nitrophenyl- and 4-nitrophenyl-tritylethers has been known for long time [141], however, no rate data or results of sensitization or quenching experiments have become available.

c) A case of photo-denitration of 4-chloronitrobenzene has been reported to occur in aqueous methanolic sodium nitrite solution [142].

d) 1-Nitronaphthalenes are converted into 1-chloronaphthalenes upon irradiation in several alkyl chlorides or mixtures of hydrochloric acid with chloroform, carbon tetrachloride, or acetic acid [143].

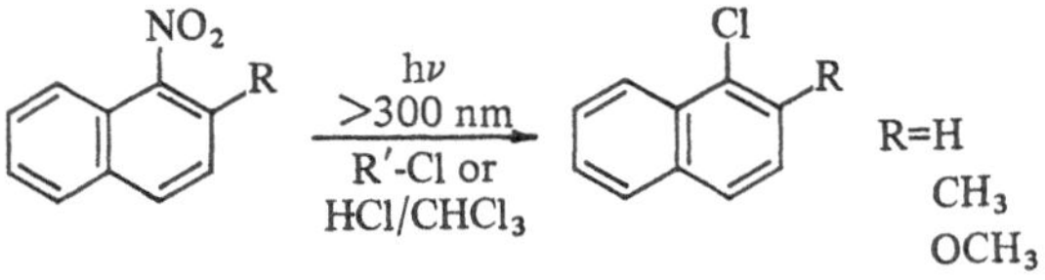

Under spectroscopic conditions, a 100% conversion is easily reached and the reaction is reported to be remarkably clean. High yields are obtained in preparative runs. In the same way both nitro groups are exchanged for chlorine in 1,5-dinitronaphthalene, whereas 1,8-dinitronaphthalene gives rise to a trichloronaphthalene. 2-Nitronaphthalene was unreactive.

3. Decarboxylations

Photodecarboxylations of 3-nitrophenyl acetates and 4-nitrophenyl acetates [144], α-(2,4-dinitrophenyloxy)-acetic acids [145], α-(2-nitrophenylthio)-acetic acids [146], N-(4-nitrophenyl)-amino acids [147], N-(2,4-dinitrophenyl)-amino acids [148] and 3-nitro-2-pyridyl amino acids [149] have been of considerable interest. A generally applicable mechanistic scheme does not exist so far, and except for 3-nitrophenyl acetate [144] in *ortho* and/or *para* position of the phenyl (pyridyl) ring seems to be a prerequisite for efficient decarboxylation.

Photodecarboxylation of similar but nitro group free substrates have also been verified, however. N-2-chlorophenylglycine and (although less efficient) α-phenylthio-acetic acid have been decarboxylated by irradiation in acetonitrile solution in the presence of l-nitronaphthalene, 4-nitro-biphenyl, 1,4-dinitro-benzene, 4-nitrotoluene and nitrobenzene [150].

It is assumed that an excited state charge transfer complex is formed between the nitroaromatic in its first triplet state and the respective substrate. Internal proton transfer is immediately followed by liberation of carbon dioxide. Finally hydrolysis of the hemiacetal $Ar'-X-CH_2OH$ ($X = NH$ or S) leads to 2-chloro-aniline or thiophenol, respectively. In the decarboxylation of α-phenylthio-acetic acid, some methyl-phenylsulfide is also formed. (π,π^*)-nitroaromatics are more reactive than nitro compounds with lowest (n,π^*)-triplets [150].

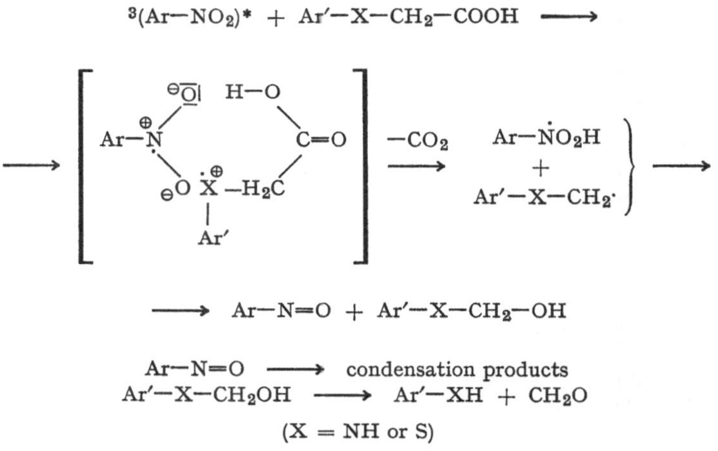

4. Miscellaneous Reactions

There are many reports on other intriguing light-induced reactions of aromatic nitro compounds, such as:

Preferential reduction of a nitro group in the presence of a carbonyl group in 4-nitroacetophenone [151], intramolecular rearrangements of o-nitro-benzanilides [152], intramolecular cyclizations of o-nitro-*tert*-anilines to benzimidazol-1-oxides [153,154], cyclizations of acylated 2-nitrodiphenylamines to phenazine-1-oxides [155], intramolecular additions of nitro groups to double bonds [156], remarkably efficient ring openings of 1-(2,4-dinitrophenyl)pyrazoles [157], oxygen transfer from the nitro group to sulfur in 2-nitro-diphenylsulfoxides [158], generation of a sulfenium ion (or its precursor) by irradiation of 2,4-dinitrophenylsulfenylacetate [159], liberation of carboxylic acids from N-alkyl-N-acyl-2-nitroanilines [160], and applications of o-nitrobenzyl derivatives [161–163] or o-nitrobenzylidene derivatives [164] as light-sensitive protecting groups for sugars [161,164], amino acids [162], and carboxylic acids [162] or in solid phase peptide synthesis [163].

While no experiments directed towards elucidation of the multiplicity or configuration of the reacting excited states have been reported, it should be pointed out that the work in question certainly had strictly chemical aims. But the reactions listed above nicely supplement what had previously been discussed in detail and show the variety of reactions encountered in the photochemistry of nitroaromatics.

Acknowledgement. Generous support of the work carried out in the author's laboratory by Deutsche Forschungsgemeinschaft, Fonds der Chemischen Industrie and BASF Aktiengesellschaft is gratefully acknowledged. The author expresses thanks to Drs. A. C. Testa, St. Johns University, New York, and J. Cornelisse, Rijksuniversiteit Leiden, for critical comments on the manuscript.

References

[1] McGlynn, S. P., Azumi, T., Kinoshita, M.: Molecular spectroscopy of the triplet state, p. 251 ff. Englewood Cliffs: Prentice Hall 1969.

[2] Khalil, O. S., Bach, H. G., McGlynn, S. P.: J. Mol. Spectry. *35*, 455 (1970).

[3] a) Rusakowicz, R., Testa, A. C.: Spectrochim. Acta *27 A*, 787 (1971);
b) Ref. [68];
c) Ref. [102].

[4] Brinen, J. S., Singh, B.: J. Am. Chem. Soc. *93*, 6623 (1971).

[5] Lim, E. C., Stanislaus, J.: Chem. Phys. Letters *6*, 195 (1970).

[6] Kanamaru, N., Okajima, S., Kimura, K.: Bull. Chem. Soc. Japan *45*, 1273 (1972).

[7] Rao, C. N. R.: Spectroscopy of the nitro group. Chapter 2 in Ref. [9a].

[8] Morrison, H. A.: The photochemistry of the nitro and nitroso groups. Chapter 4 in Ref. [9a].

[9] Wagniere, G. H.: Theoretical aspects of the C—NO and C—NO$_2$ bonds. Chapter 1 in Ref. [9a].

[9a] Feuer, H. (ed.): The chemistry of the nitro and nitroso groups, Part 1. New York: Interscience Publishers 1969.

[10] Ciamician, G., Silber, P.: Ber. Deut. Chem. Ges. *19*, 2889 (1886); *38*, 3813 (1905).

[11] Hurley, R., Testa, A. C.: J. Am. Chem. Soc. *88*, 4330 (1966).

[12] Hurley, R., Testa, A. C.: J. Am. Chem. Soc. *89*, 6917 (1967).

[13] Hurley, R., Testa, A. C.: J. Am. Chem. Soc. *90*, 1949 (1968).

[14] Trotter, W., Testa, A. C.: J. Am. Chem. Soc. *90*, 7044 (1968).

[15] Cu, A., Testa, A. C.: J. Am. Chem. Soc. *96*, 1963 (1974).

[16] Barltrop, J. A., Bunce, N. J.: J. Chem. Soc. (C) *1968*, 1467.

[17] Hashimoto, S., Kano, K.: Bull. Chem. Soc. Japan *45*, 549 (1972).

[18] Wubbels, G. G., Jordan, J. W., Mulls, N. S.: J. Am. Chem. Soc. *95*, 1281 (1973).

[19] Jaffé, H. H., Orchin, M.: Theory and applications of ultraviolet spectroscopy, p. 183. New York: J. Wiley and Sons 1962.

[20] Nagakura, S., Kojima, M., Maruyama, Y.: Mol. Spectry. *13*, 174 (1964).

[21] Janzen, E. G., Gerlock, J. L.: J. Am. Chem. Soc. *91*, 3108 (1969).

[22] McMillan, M., Norman, R. O. C.: J. Chem. Soc. (B) *1968*, 590.

23) a) Cowley, D. J., Sutcliffe, L. H.: J. Chem. Soc. (B) *1970*, 569;
   b) Sleight, R. B., Sutcliffe, L. H.: Trans. Faraday Soc. *67*, 2195 (1971).
24) Wong, S. K., Wan, J. K. S.: Can. J. Chem. *51*, 753 (1973).
25) Ward, R. L.: J. Chem. Phys. *38*, 2588 (1963).
26) Chachaty, C., Forchioni, A.: Tetrahedron Letters *1968*, 307.
27) Brown, J. K., Williams, W. G.: Chem. Commun. *1966*, 495.
28) Lamola, A. A., Hammond, G. S.: J. Chem. Phys. *43*, 2129 (1965).
29) Lewis, G. N., Kasha, M.: J. Am. Chem. Soc. *66*, 2100 (1944).
30) Charlton, J. L., Liao, C. C., de Mayo, P.: J. Am. Chem. Soc. *93*, 2463 (1971).
31) Moore, W. M., Hammond, G. S., Foss, R. P.: J. Am. Chem. Soc. *83*, 2789 (1961).
32) Weller, J. W., Hamilton, G. A.: J. Chem. Soc. (D) *1970*, 1390.
33) Scholl, P. C., Van de Mark, M. R.: J. Org. Chem. *38*, 2376 (1973).
34) Bamberger, E., Busdorf, H., Szolaysky, B.: Ber. Deut. Chem. Ges. *32*, 210 (1899).
35) Letsinger, R. L., Wubbels, G. G.: J. Am. Chem. Soc. *88*, 5041 (1966).
36) Schulman, S. G., Sanders, L. B., Winefordner, J. D.: Photochem. Photobiol. *13*, 381 (1971).
37) Seely, G. R.: J. Phys. Chem. *73*, 117 (1969).
38) Petersen, W. C., Letsinger, R. L.: Tetrahedron Letters *1971*, 2197.
39) Döpp, D., Müller, D., Sailer, K.-H.: Tetrahedron Letters *1974*, 2137.
40) Vink, J. A. J., Cornelisse, J., Havinga, E.: Rec. Trav. Chim. *90*, 1333 (1971).
41) Trotter, W., Testa, A. C.: J. Phys. Chem. *75*, 2415 (1971).
42) Kitaura, Y., Matsuura, T.: Tetrahedron *27*, 1583 (1971).
43) a) El'tsov, A. V., Kuznetsova, N. A., Frolov, A. N.: Zh. Organ. Khim. *7*, 817 (1971);
   b) Frolov, A. N., Kuznetsova, N. A., El'tsov, A. V., Rtishchev, N. I.: Zh. Organ. Khim.
   9, 963 (1973).
44) Frolov, A. N., El'tsov, A. V., Sosonkin, I. M., Kuznetsova, N. A.: Zh. Organ. Khim. *9*,
   973 (1973).
45) Jones, L., Kudrna, J., Foster, J.: Tetrahedron Letters *1969*, 3263.
46) Roth, H. J., Adomeit, M.: Tetrahedron Letters *1969*, 3201.
47) Hoshino, O., Sawaki, S., Miyazaki, N., Umezawa, B.: J. Chem. Soc. (D) *1971*, 1572.
48) Döpp, D.: Chem. Ber. *104*, 1043 (1971).
49) Takami, M., Matsuura, T., Saito, I.: Tetrahedron Letters 1974, 661.
50) a) Hart, H., Link, J. W.: J. Org. Chem. *34*, 758 (1969);
   b) Reid, S. T., Tucker, J. N.: Chem. Commun. *1970*, 1286.
51) Bakke, J.: Acta Chem. Scand. *24*, 2650 (1970).
52) Barclay, L. R. C., McMaster, I. T.: Can. J. Chem. *49*, 676 (1971).
53) Döpp, D.: Chem. Commun. *1968*, 1284.
54) Döpp, D.: Chem. Ber. *104*, 1035 (1971).
55) Döpp, D.: Tetrahedron Letters *1971*, 2757.
56) Döpp, D., Sailer, K.-H.: a) Chem. Ber., in press;
   b) Tetrahedron Letters *1971*, 2761.
57) Döpp, D., Brugger, E.: Chem. Ber. *106*, 2166 (1973).
58) Döpp, D., Sailer, K.-H.: In preparation.
59) Döpp, D., Brugger, E.: In preparation.
60) Döpp, D.: Unpublished.
61) Döpp, D.: Tetrahedron Letters *1972*, 3215.
62) Döpp, D.: Chem. Ber. *104*, 1058 (1971).
63) Büchi, G., Ayer, D. E.: J. Am. Chem. Soc. *78*, 689 (1956).
64) Charlton, J. L., de Mayo, P.: Can. J. Chem. *46*, 1041 (1969). (Full paper: Ref. 30)).
65) Saito, I., Takami, M., Matsuura, T.: Chem. Letters *1972*, 1195.
66) Saito, I., Takami, M., Matsuura, T.: Tetrahedron Letters *1974*, 659.
67) Saito, I., Takami, M., Matsuura, T.: Manuscripts of Contributed papers, p. 134, V. IUPAC-
   Symposium on Photochemistry, Enschede, 1974.
68) Lammers, J. G., de Gunst, G. P., Havinga, E.: Rec. Trav. Chim. *92*, 1386 (1973).
69) McClure, D. S.: J. Chem. Phys. *17*, 905 (1949).
70) Corkill, J. M., Graham-Bryce, I. J.: J. Chem. Soc. *1961*, 3893.
71) Mikula, J. J., Anderson, R. W., Harris, L. E., Stuebing, E. W.: J. Mol. Spectry. *42*, 350
   (1972).

72) Trotter, J.: Acta Cryst. *13*, 95 (1960).
73) Hammond, G. S., Leermakers, P. A.: J. Am. Chem. Soc. *84*, 207 (1962).
74) Trotter, W., Testa, A. C.: J. Phys. Chem. *74*, 845 (1970).
75) Hashimoto, S., Kano, K.: Kogyo Kagaku Zasshi, *72* 188 (1969); C. A. *71*, 2765 t.
76) Obi, K., Bottenheim, J. W., Tanaka, I.: Bull. Chem. Soc. Japan *46*, 1060 (1973).
77) Hashimoto, S., Kano, K., Ueda, K.: Tetrahedron Letters *1969*, 2733; Bull. Chem. Soc. Japan *44*, 1102 (1971).
78) Cu, A., Testa, A. C.: J. Phys. Chem. *77*, 1487 (1973).
79) Kaneko, C., Yamada, S., Yokoe, I., Hata, N., Ubukata, Y.: Tetrahedron Letters *1966*, 4729. — Kaneko, C., Yamada, S., Yokoe, I.: Chem. Pharm. Bull. (Tokyo) *15*, 535 (1967).
80) Kaneko, C., Yokoe, I., Yamada, S.: Tetrahedron Letters *1967*, 775.
81) See for example: Chapman, O. L., Griswold, A. A., Hoganson, E., Lenz, G., Reasoner, J.: Pure Appl. Chem. *9*, 585 (1964).
82) Hata, N., Okutsu, E., Tanaka, I.: Bull. Chem. Soc. Japan *41*, 1769 (1968).
83) Hata, N., Ono, I., Tsuchiya, T.: Bull. Chem. Soc. Japan *45*, 2386 (1972).
84) Ono, I., Hata, N.: Bull. Chem. Soc. Japan *45*, 2951 (1972).
85) Footnote 8 in Ref.[84].
86) For a recent review, see: Cornelisse, J.: Pure Appl. Chem., in press.
87) Review (orientation rules): Lok, C. M., Havinga, E.: Koninkl. Nederl. Akad. Wetensch. Amsterdam, Proceedings Ser. B *77*, 15 (1974).
88) Review: Havinga, E., de Jongh, R. O., Kronenberg, M. E.: Helv. Chim. Acta *50*, 2550 (1967).
89) Review: Havinga, E., Kronenberg, M. E.: Pure Appl. Chem. *16*, 137 (1968).
90) Havinga, E., de Jongh, R. O., Dorst, W.: Rec. Trav. Chim. *75*, 378 (1956).
91) Havinga, E., de Jongh, R. O.: Bull. Soc. Chim. Belg. *71*, 803 (1962).
92) De Vries, S., Havinga, E.: Rec. Trav. Chim. *84*, 601 (1965).
93) De Jongh, R. O., Havinga, E.: Rec. Trav. Chim. *85*, 275 (1966).
94) Kronenberg, M. E., van der Heyden, A., Havinga, E.: Rec. Trav. Chim. *85*, 56 (1966).
95) Nijhoff, D. F., Havinga, E.: Tetrahedron Letters *1965*, 4199.
96) Cornelisse, J., Havinga, E.: Tetrahedron Letters *1966*, 1609.
97) Van Vliet, A., Kronenberg, M. E., Havinga, E.: Tetrahedron Letters *1966*, 5957.
98) Van der Stegen, G. H. D., Poziomek, E. J., Kronenberg, M. E., Havinga, E.: Tetrahedron Letters *1966*, 6371.
99) Kronenberg, M. E., van der Heyden, A., Havinga, E.: Rec. Trav. Chim. *86*, 254 (1967).
100) Van Vliet, A., Cornelisse, J., Havinga, E.: Rec. Trav. Chim. *88*, 1339 (1969).
101) Van Vliet, A., Kronenberg, M. E., Cornelisse, J., Havinga, E.: Tetrahedron *26*, 1061 (1970).
102) Beijersbergen van Henegouwen, G. M. J., Havinga, E.: Rec. Trav. Chim. *89*, 907 (1970).
103) Hartsuiker, J., de Vries, S., Cornelisse, J., Havinga, E.: Rec. Trav. Chim. *90*, 611 (1971).
104) Lok, C. M., Lugtenburg, J., Cornelisse, J., Havinga, E.: Tetrahedron Letters *1970*, 4701. Lok, C. M., den Boer, M. E., Cornelisse, J., Havinga, E.: Tetrahedron *29*, 867 (1973).
105) Brasem, P., Lammers, J. G., Cornelisse, J., Lugtenburg, J., Havinga, E.: Tetrahedron Letters *1972*, 685.
106) Vink, J. A. J., Verheijdt, P. L., Cornelisse, J., Havinga, E.: Tetrahedron *28*, 5081 (1972).
107) De Gunst, G. P., Havinga, E.: Tetrahedron *29*, 2167 (1973).
108) De Gunst, G. P.: Thesis, Leiden, 1971.
109) Den Heijer, J., Spee, T., de Gunst, G. P., Cornelisse, J.: Tetrahedron Letters *1973*, 1261.
110) Lammers, J. G., Lugtenburg, J.: Tetrahedron Letters *1973*, 1777.
111) Johnson, R. M., Rees, C. W.: Proc. Chem. Soc. *1964*, 213.
112) Gold, C., Rochester, C. H.: Proc. Chem. Soc. *1960*, 403; J. Chem. Soc. *1964*, 1717.
113) Letsinger, R. L., Ramsay, O. B.: J. Am. Chem. Soc. *86*, 1447 (1964).
114) Letsinger, R. L., Ramsay, O. B., McCain, J. H.: J. Am. Chem. Soc. *87*, 2945 (1965).
115) Letsinger, R. L., McCain, J. H.: J. Am. Chem. Soc. *88*, 2884 (1966).
116) Letsinger, R. L., Hautala, R. R.: Tetrahedron Letters *1969*, 4205.
117) Letsinger, R. L., Steller, K. E.: Tetrahedron Letters *1969*, 1401.
118) Letsinger, R. L., McCain, J. H.: J. Am. Chem. Soc. *91*, 6425 (1969).
119) Steller, K. E., Letsinger, R. L.: J. Org. Chem. *35*, 308 (1970).

120) Hautala, R. R., Letsinger, R. L.: J. Org. Chem. *36*, 3762 (1971).
121) Petersen, W. C., Letsinger. R. L.: Tetrahedron Letters *1971*, 2197.
122) Párkányi, C., Gutiérrez, A. R., Lee, Y. J., Lee, S. A.: IV. IUPAC-Symposium on Photo-chemistry, Baden-Baden 1972, Manuscripts of Contributed Papers, p. 180.
123) Groen, M. B., Havinga, E.: Mol. Photochem. *6*, 9 (1974).
124) Fráter, C., Cornelisse, J.: In: Srinivasan, R., Roberts, T. D.: Organic Photochemical Syntheses, *1*, p. 74. New York: Wiley-Interscience 1971.
125) Vink, J. A. J., Lok, C. M., Cornelisse, J., Havinga, E.: J. Chem. Soc. Chem. Commun. *1972*, 710.
126) Vink, J. A. J.: Thesis. Leiden 1972.
127) Nilsson, S.: Acta Chem. Scand. *27*, 329 (1973).
128) Ullmann, E., Singh, P.: J. Am. Chem. Soc. *94*, 5077 (1972).
129) Beiersbergen van Henegouwen, G. M. J.: Thesis, Leiden, 1970.
130) Lammers, J. G.: Thesis. Leiden, 1974.
131) Powers, L. J.: J. Pharm. Sci. *60*, 1425 (1971).
132) Lopp, J. G., Hendren, R. W., Wildes, P. D., Whitten, D. G.: J. Am. Chem. Soc. *92*, 6440 (1970).
133) Bent, D. V., Schulte-Frohlinde, D.: J. Phys. Chem. *78*, 446 (1974).
134) Hammond, G. S., Saltiel, J., Lamola, A. A., Turro, N. J., Bradshaw, J. B., Cowan, D. O., Coursell, R. C., Vogt, V., Dalton, C.: J. Am. Chem. Soc. *86*, 3197 (1964).
135) Bent, D. V., Schulte-Frohlinde, D.: J. Phys. Chem. *78*, 451 (1974).
136) Jungmann, H., Güsten, H., Schulte-Frohlinde, D.: Chem. Ber. *101*, 2690 (1968).
137) Hunt, R., Reid, S. T., Chem. Commun, *1970*, 1576, J. Chem. Soc. Perkin I *1972*, 2527.
138) Bamberger, E., Elger, F.: Liebigs Ann. Chem. *475*, 288 (1928).
139) Wubbels, G. W., Hautala, R. R., Letsinger, R. L.: Tetrahedron Letters *1970*, 1689. See also: Reisch, J., Weidmann, K.G.: Arch. Pharmaz. *304*, 906 (1971).
140) De Bie, D. A., Havinga, E.: Tetrahedron *21*, 2359 (1965).
141) Zimmermann, H. E., Somasekhara, S.: J. Am. Chem. Soc. *85*, 922 (1963).
142) Frolov, A. N., El'tsov, A. V.: Zh. Organ. Khim. *6*, 637 (1970).
143) Fráter, G., Havinga, E.: Tetrahedron Letters *1969*, 4603; Rec. Trav. Chim. *89*, 273 (1970).
144) Margerum, J. D., Petrusis, C. T.: J. Am. Chem. Soc. *91*, 2467 (1969).
145) McFarlane, P. H., Russel, D. W.: Chem. Commun. *1969*, 475.
146) Goudie, R. S., Preston, P. N.: J. Chem. Soc. (C) *1971*, 3081.
147) McFarlane, P. H., Russel, D. W.: Tetrahedron Letters *1971*, 725.
148) Neadle, D. J., Pollitt, R. J.: J. Chem. Soc. (C) *1967*, 1764; *1969*, 2127.
149) Bordignon, E., Aloisi, G. G., Signor, A.: Gazz. Chim. Ital. *100*, 802 (1970). — Aloisi, G. G., Bordignon, E., Signor, A.: J. Chem. Soc. Perkin II *1972*, 2218.
150) Davidson, R. S., Korkut, S., Steiner, P. R.: Chem. Commun. *1971*, 1052.
151) Blossey, E. C., Corley, A.: J. Chem. Soc. Chem. Commun. *1972*, 895.
152) Gunn, B. C., Stevens, M. F. G.: J. Chem. Soc. Perkin I *1973*, 1682; J. Chem. Soc. Chem. Commun. *1972*, 835.
153) Fielden, R., Meth-Cohn, O., Suschitzky, H.: J. Chem. Soc. Perkin I *1973*, 696.
154) Heine, H. W., Blosick, G. J., Lowrie, G. B.: Tetrahedron Letters *1968*, 4801.
155) Maki, Y., Suzuki, M., Hosokami, T., Furuta, T.: J. Chem. Soc. Perkin I *1974*, 1354.
156) Van Allen, J. A., Farid, S., Reynolds, G. A., Chie Chang, S.: J. Org. Chem. *38*, 2834 (1973).
157) Bouchet, P., Coquelot, C., Elguero, J. Jacquier, R.: Tetrahedron Letters *1973*, 891.
158) Tanikaga, R., Higashio, Y., Kaji, A.: Tetrahedron Letters *1970*, 3273.
159) Barton, D. H. R., Nakano, T., Sammes, P. G.: J. Chem. Soc. (C) *1968*, 322.
160) Amit, B., Patchornik, A.: Tetrahedron Letters *1973*, 2205.
161) Zehavi, U., Patchornik, A.: J. Org. Chem. *37*, 2281, 2285, (1972).
162) Patchornik, A., Amit, B., Woodward, R. B.: J. Am. Chem. Soc. *92*, 6333 (1970).
163) Rich, D. H., Gurawa, S. K.: J. Chem. Soc. Chem. Commun. *1973*, 610.
164) Collins, P. M., Oparaeche, N. N.: J. Chem. Soc. Chem. Commun. *1972*, 532.

Received September 5, 1974

Triplet-Intermediates from Diazo-Compounds (Carbenes)

Prof. Dr. Heinz Dürr

Institut für Organische Chemie der Universität, Saarbrücken

Contents

1. Introduction

The photolysis or thermolysis of diazo compounds results in the formation of carbenes. These reactions were identified as early as 1901 by Hantzsch and Lehmann [1] and Staudinger and Kupfer [2], who decomposed diazomethane photochemically and thermally, respectively. The subsequent work of Hine [3] and Doering [4] started the era of carbene chemistry. Excellent reviews of the chemistry of carbenes are available [5-27].

Photolysis: In photochemical generation of a carbene, excitation of the diazo compound must occur. The absorption of diazo compounds between $400 - 500$ nm, with a low extinction coefficient ($\varepsilon \sim 10$), has been assigned to a forbidden $n \rightarrow \pi^*$ transition [28, 29]. The fine structure of this band indicates that the photoexcited state is bonding. On the other hand the photoelectron spectrum of diazo-isopropane shows that the lowest energy ionization band is most probably due to a π-band [30] (b_2-π-Orbital). More work is needed to clarify this point.

Photoexcitation of the long wave bands of diazo compounds results in an excited singlet state of the diazo compound, which may then react as follows:
1. undergo intersystem crossing to the triplet diazo compound,
2. go back to the starting compound by internal conversion in a radiationless decay process,
3. decompose to a singlet carbene.

The singlet carbene can undergo intersystem crossing to the triplet carbene; the latter on the other hand, can also be formed by direct decomposition of the excited triplet state of the diazo compound. These possibilities are shown in the following scheme:

The rate of disappearance of the diazo compound, *e.g.* the quantum yield of the photolysis, should be independent of the nature and concentration of the reactants. This is the kinetic criterion for the intermediacy of a *free carbene*. The quantum yield data of diazo compound photolyses are shown in Table 1.

Carbenes are named in the usual way by using
a) the carbinol convention: $CH_3\ddot{C}H = $ methylcarbene or
b) radicals according to the IUPAC rules: $CH_3\ddot{C}H = $ ethylidene.
The term "methylene" is exclusively reserved for: $\ddot{C}H_2$. Spectroscopic investigations demonstrated that photolysis of diazo compounds in fact produces free carbenes. A flash photolysis of diazomethane gave methylene whose spectrum could be recorded [31]. ESR-spectra were taken of a series of triplet carbenes which had been obtained by direct irradiation of diazo compounds in various matrices at low temperatures (see p. 97).

Table 1. UV-absorptions and quantum yields of decomposition of some diazocompounds

	λmax (nm)	ε	Quantum yield $\Phi^{1)}$ for decomposition of $R_2C{=}N_2$	Ref.
CH_2N_2	ca. 410	3	4	29 a)
$CH_3{-}CHN_2$	440	3.5	—	29 a)
	470	3.5		
$(C_6H_5)_2{-}CN_2$	526	101	—	29 b)
	288	21300	0.78	
(naphthalene-OCH_3, ${=}N_2$)	513	103	—	29 b)
	287	20600	0.69	
$C_2H_5OOC{-}CHN_2$	360	21	—	29 b)
	269	7110	0.66	
	247	7650	—	
$C_6H_5CO{-}CHN_2$	294	13500	0.46	29 b)
	250	12300	—	
$(C_6H_5CO)_2{-}CN_2$	275	16600	0.31	29 b)
	256	22200	—	
$C_6H_5CO{-}CN_2{-}CO_2CH_3$	274	10100	0.35	29 b)
	253	10700	—	
(naphthalene-$COCH{=}N_2$)	301	11500	0.31	29 b)
(thiophene-$COCH{=}N_2$)	309	19700	0.36	29 b)
	261	9050		
(bicyclic ketone N_2)	301	3310	0.24	29 b)
	252	12000	—	
(indanone ${=}N_2$)	321	12150	0.14	29 b)
	256	19000	—	
(tetralone N_2)	326	12000	0.21	29 b)
	260	9780	—	
$N_2CH{-}CO{-}(CH_2)_4{-}COCHN_2$	270	17300	0.34	29 b)
	248	19000	—	
$N_2CH{-}CO{-}(C_6H_4){-}COCHN_2$	311	22500	0.15	29 b)
	264	17500	—	
$N_2CH{-}CO{-}CO{-}CHN_2$	318	11700	0.31	29 b)
	270	15900	—	

[1]) Quantum yields Φ were determined in methanol at the wavelenght shown in the table.

Isotopically labelled $^{15}N_2$ could be incorporated in partially decomposed diazomethane using solid or gaseous $^{15}N_2$ [32, 33]. The detailed route to carbenes and the electronic configuration of the triplet state of carbenes will be discussed below. Several other photochemical reactions also yield diazo compounds as intermediates. These reactions are summarized in the following scheme, but will not be further dealt with in this article.

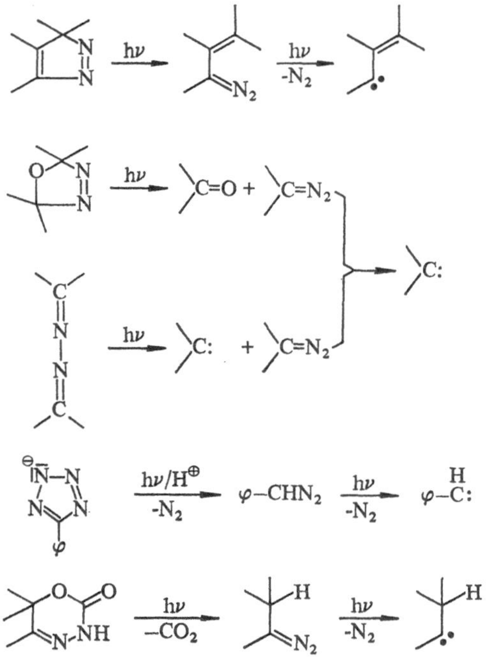

Thermolysis: The thermolysis of diazomethane was studied by Staudinger [2] a long time ago. In the presence of CO some ketene was formed most probably by the reaction between methylene and CO.

Kinetic studies of the thermolysis of diazomethane were carried out by various authors. These experiments demonstrated that the decomposition of diazomethane was a first order reaction [41–43]. Similar investigations of the pyrolysis of diphenyl-diazomethane in xylene or 1-methylnaphthalene also showed that the disappearance of diphenyl-diazomethane is a first order process. It may be concluded that a free carbene is involved in these reactions, in accordance with the following scheme [44, 45].

$$\varphi_2C{=}N_2 \xrightarrow{k_1} \varphi_2C: + N_2 \xrightarrow[k_2]{\varphi_2C=N_2} \varphi_2C{=}N{-}N{=}C\varphi_2$$

$$\varphi_2C: \xrightarrow{k_3} \text{products}$$

When phenyl-diazomethane was employed, the observed order of reactions was intermediate between first and second [46].

However, acid catalysis can interfere in thermal reactions of diazoalkanes and must always be taken into consideration.

To summarize, it may be said that, broadly speaking, *free carbenes* are formed from diazo-compounds in photochemical and thermal reactions, whereas all other methods yield *carbenoids*.

2. Electronic States and Structure of Triplet Carbenes

2.1. Calculations

In this section the electronic states of carbenes will be briefly discussed. As an example, some semiempirical calculations are presented for methylene, in order to gain a better understanding of the electronic states of carbenes.

Carbenes with two orbitals, occupied by one electron each, have a total spin number of $s = 1$; the multiplicity is therefore $2s + 1 = 3$, *i.e.*, a triplet state. This is the usual ground state of a carbene. If however, the two orbitals are not degenerate, the multiplicity can become 1, *i.e.*, a singlet state is possible. Methylene — and other carbenes — can, in principle, be linear or bent. The binding is achieved by using the 15 hydrogen $1s$ and the $2s$ and $2p$ carbon orbitals.

In the linear methylene ($D_{\infty h}$-symmetry) binding results from the overlapping of these orbitals, to give the 2 σ_g and 1 σ_u-orbitals. The two remaining π_u-orbitals ($2p_x$ and $2p_y$) are degenerate. Linear methylene should be a triplet $^3\Sigma_g^-$; this follows from the simple MO-model.

linear: $D_{\infty h}-$ bent: C_{2v}

Reducing the angle of 180° between the two hydrogens is without effect on the out-of-plane orbital $p_x \equiv \pi_{uy}$. On the other hand, $\pi_{ux} \equiv \sigma$ acquires more s-character and is thus stabilized. This stabilization terminates the degeneracy of the two orbitals, when a methylene with an angle of 90° and a singlet ground state [18b] is obtained (see Fig. 1).

This simple consideration led to a number of highly sophisticated quantum-mechanical calculations for methylene, such as extended Hückel, MINDO/2 and 3, and ab initio calculations. The results of these calculations are presented in Table 3.

The calculations show that the ground state of methylene should be the $^3B_1$ triplet state (see Table 2) with a bond angle of 132—134°. This is clearly borne out by the more sophisticated calculations such as the MINDO/2, MINDO/3 and

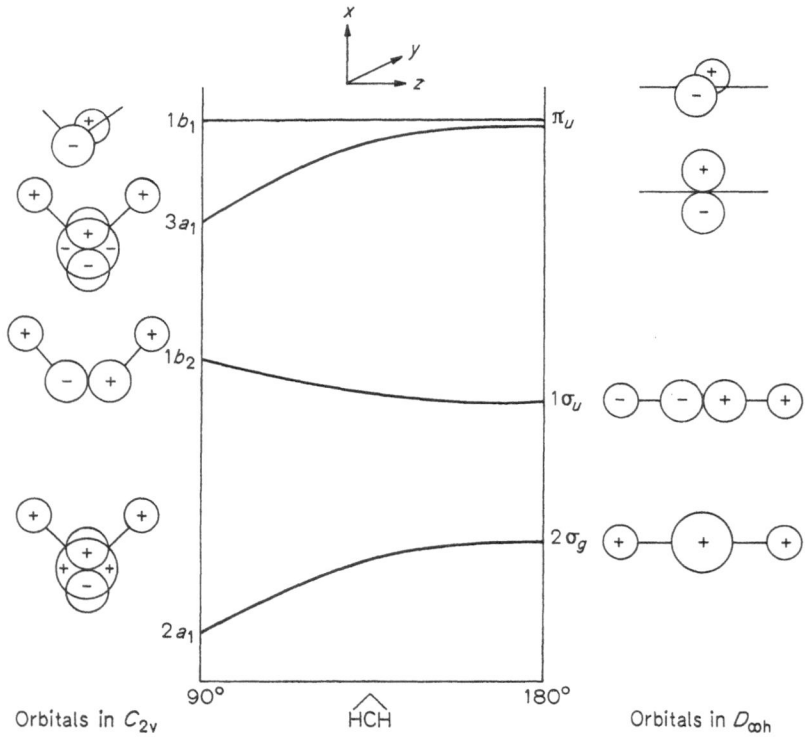

Fig. 1. Molecular orbitals of methylene and their energy as a function of bond angle

Table 2. Electronic states of methylene

State	Geometry	Term symbol	Occupied orbitals		
			Carbon 1_s	C—H bonds	Non-bonding
Lowest triplet	Bent (C_{2v})	$^3B_1$	$(1a_1)^2$	$(2a_1)^2(1b_2)^2$	$(3a_1)(1b_1)$
	Linear $(D_{\infty h})$	$^3\sum_g^-$	$(1\sigma_g)^2$	$(2\sigma_g)^2(1\sigma_u)^2$	$(\pi_{ux})(\pi_{uy})$
Lowest singlet	Bent (C_{2v})	$^1A_1$	$(1a_1)^2$	$(2a_1)^2(1b_2)^2$	$(3a_1)^2$
First excited singlet	Bent (C_{2v})	$^1B_1$	$(1a_1)^2$	$(2a_1)^2(1b_2)^2$	$(3a_1)(1b_1)$

the ab initio calculations, which are in good agreement with the most recent experimental values. Figure 2 shows, however, that the minimum of the $^3B_1$ state is very shallow, so that only a small amount of energy is necessary for the angle of the methylene triplet state to become slightly bent.

The energy gap between the triplet ground state $^3B_1$ and the $^1A_1$ singlet state has been calculated by MINDO/2 and MINDO/3 [52] to be 28.3 and 8.7 kcal/mole, respectively. The latter value is quite close to the experimental value of about 8 kcal/mole [54, 55].

Table 3. Calculated geometries and energies of the lowest electronic states of methylene[1])

	CI[47]	Ab initio VB[48]	EHMO[50]	SCF CI[51]	MINDO/2[52]	MINDO/3[52]	Ab initio-SCF[48,49,53]	Exper. values
r_{C-H}(Å)	1.11 1.17	—	—	1.096	1.062	1.078	1.062	1.078 [77]
$^3B_1$ HCH[3] (T_1)	129°	138°	155°	135°	142°	134.1°	132.5° [48,49]	136° [60,76,77]
ΔH[2]	—	—	—	—	67.5	91.5	—	—
r_{C-H}	—	—	—	—	1.097	1.12	1.100	1.11 [76]
$^1A_1$ HCH[3] (S_0)	90°	108°	115°	—	107°	100.2°	105° [49]	102.4° [76,77]
ΔH[2]	—	—	—	—	95.8	100.2	—	91.9; 86±6; 94.6—94.5 [53a,76]
r_{C-H}	—	—	—	—	1.050	1.078	1.092 [51]	—
$^1B_1$ HCH[3] (S_1)	132°	148°	—	—	180°	141.7°	143.8° [51]	140±15° [76,77]
ΔH[2]	—	—	—	—	97.0	125.0	—	—

1) Abbreviations: CI, configuration interaction; VB, valence bond; EHMO, extended Hückel molecular orbital; SCF, self-consistent field.
2) Heat of formation, 3) bond angle.

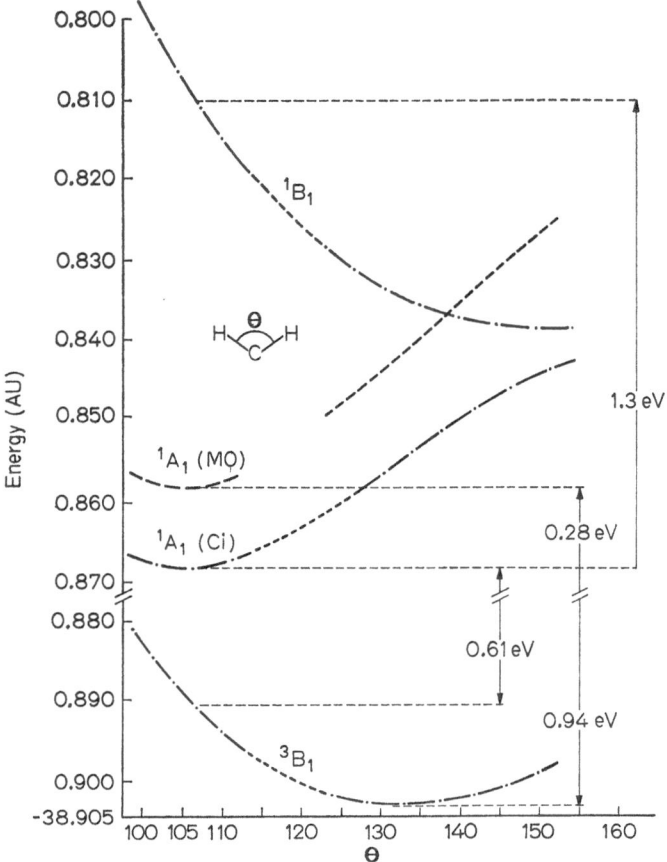

Fig. 2. Energies of various electronic states of CH_2 [23]

Generalization from Methylene to Other Carbenes

It has been seen in the case of methylene that the degeneracy of the σ- and p-orbital of a carbene can be destroyed in the following manners:

1. by bending; this stabilizes the σ-orbital by intensifying its s-character.

2a. the degeneracy can be broken by connecting the carbene center with substituents having low-lying vacant π^*-orbitals. The approach must be such that the empty π^*-orbitals interact selectively with one orbital. The interaction diagram is shown in Fig. 3.

In order to have a truly large σ, p separation which can then lead to a *singlet ground state* carbene, the energy gap produced must be larger than the electron pairing energy (about 1 eV = 23 kcal/mole). Extended Hückel calculations [50, 56] show, however, that this interaction rarely gives rise to a sufficiently large energy gap. The stabilization achieved by this orbital interaction viz. lowering the p-orbital of the carbene, proceeds in the same direction as the stabilization of the σ orbital, so that no large net effect is obtained. A carbene of that type would be formyl-carbene ($H\ddot{C}-CHO$) or nitro-carbene (O_2N-C-H).

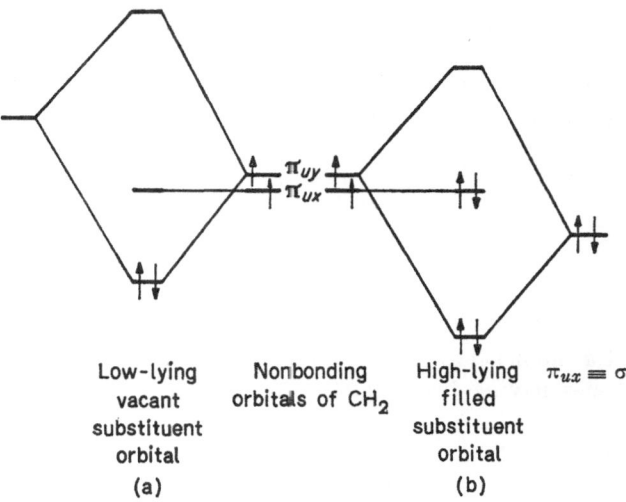

Fig. 3. Interaction diagram for orbitals capable of conjungation with the π_{uy} orbital of linear methylene [56]

An ab initio calculation for cyclopentadienylidene shows that the σp-triplet is the ground state, but the p^2-singlet is only 0.3 kcal/mole richer in energy. It follows that p^2-singlet can be easily generated and that the small energy gap should favor an equilibrium between the two states. This is, however, not observed experimentally [56a]. These calculations also predict that a cyclopentadienylidene having a pyramidal structure should be, energetically, the most stable arrangement.

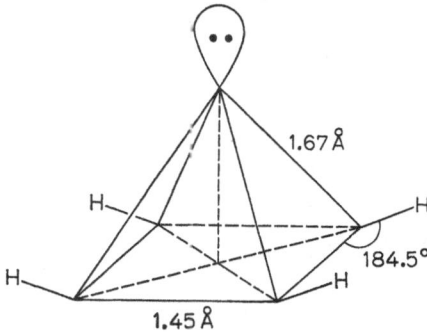

2b. The same effect can be achieved by connecting methylene to a system with high-lying occupied levels. Here the effects of bending and of the orbital interaction are opposed, and give rise to a large energy gap. Accordingly, the resulting carbene shows a large σ, p split owing to which it is the singlet configuration which is now the most stable (see Fig. 3). Carbenes of this type are difluoro-, fluoro- and chloro-carbenes for which a singlet ground state is predicted [50], in agreement with experimental data. Similar predictions were made for cyclopropenylidene [52,56], the heterocyclic carbene A [56], and vinylidene [52].

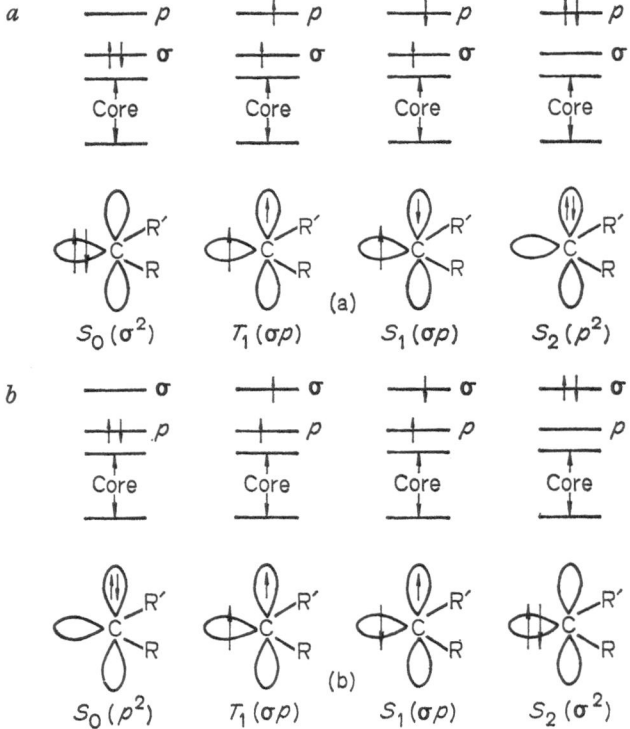

According to the nomenclature of Hoffmann, this gives rise to two different classes of carbenes having two singlet states, either σ^2 or p^2 being the lower energy state.

Fig. 4 Schematic representation of the ground and excited configurations of (a) σ^2 and (b) p^2 carbenes

Figure 4 shows that, as a matter of fact, four electronic configurations have to be considered for carbenes belonging to group 2a), viz.: the σ^2 (S_0), σp (T_1), σp (S_1) and p^2 (S_2) carbene.

Usually only the lowest-lying states S_0 and T_1 are considered. However, triplet reactions have never been interpreted as involving the excited singlet state S_1, as well; this should behave as a diradical and may show a chemistry much like that of T_1, but should be not detected in ESR-studies.

The second conclusion which may be drawn from Hoffmann's calculations [50,56] is that carbenes having substituents with vacant orbitals interacting with the p-orbital should be electrophilic, whereas the interaction with substituents with filled orbitals should result in nucleophilic behaviour. There is experimental proof [57] to the effect that singlet carbenes such as dihalocarbenes with their empty p-orbital react as electrophiles.

Similar results were obtained for cyclopentadienylidene [58]. Cyclohepta-trienylidene, which belongs to carbenes of group 2b), reacts as singlet and also clearly shows a nucleophilic character [59], in good agreement with the earlier calculations [56].

2.2. Spectroscopic Data of Triplet-Carbenes

Spectra of carbenes are very useful sources of information on the structure of the free carbenes, *e.g.* the $R-C-R$ angle, or the multiplicity of their lowest state. However, these data were mostly obtained under conditions different from those in solution, where chemical reactions normally occur. The spectra are usually recorded either in matrices at low temperatures, say at 4 or 77 °K, or in the gas phase. Only very few investigations of that type have been carried out in solution. The most important spectroscopic technique used in the investigations of carbenes is ESR. Other spectroscopic methods, such as flash photolysis which produces electronic spectra of carbenes, and infrared and lately CIDNP spectroscopy have been successfully employed.

2.2.1. Electron Spin Resonance (ESR)

This technique is most useful in studying *triplet carbenes* since it responds to triplet states only. The carbenes are mostly generated at low temperatures (4 or 77 °K) by photolysis of diazoalkanes in a) solid solutions in single crystals or b) in randomly oriented glasses.

Transitions are then possible between the three spin states $s = 1, 0$ or -1 which produce the observed ESR signals. The resulting spectrum for an organic molecule with two unpaired spins can be described by

$$\mathscr{H} = g \cdot \beta \cdot HS + DS_z^2 + E(S_x^2 - S_y^2)$$

where: S is the spin operator, g is the Landé splitting factor and D and E are the so called zero field splitting parameters, which should be present in the absence of an external field. The value of D is roughly proportional to the *separation* of the two *unpaired spins*, and is a measure of electron delocalization. E indicates the *deviation* of the spin interaction from *cylindrical symmetry*. The E/D ratio can be correlated with the $R-C-R$ bond angle in the carbenes. E/D — which is often referred to as an indication of the fractional s-character of the σ-orbital — should be zero for a linear carbene, whereas $E/D = 0.33$ would indicate an sp^2-hybridized carbene with an angle of $\sim 120°$. However, the values of E and D obtained by ESR depend on the nature of the matrices used for ESR-spectroscopy.

A survey of characteristic E- and D-values for triplet carbenes is presented in Table 4.

Table 4. Zero-field splitting parameters and derived bond angles for triplet carbenes

Carbene	D/hc (cm^{-1})	E/hc (cm^{-1})	Angle	Ref.
CH_2:	0.6844	0.0034	136°	[60]
	0.6636	<0.002		
	0.69	0.003	—	[61]
CD_2:	0.7563	0.00443	—	[60]
CF_3CH:	0.712	0.021	~160°	[62]
$CF_3CF_2CF_2CH$:	0.723	0.027	~160°	[62]
$CF_3(CF_2)_6CH$:	0.72	0.024	~160°	[62]
$(CF_3)_2C$:	0.7444	0.0437	~140°	[62]
$Ph-CH$:	0.518	0.0241	~155°	[63]
$\alpha-C_{10}H_7CH$:				
anti	0.4555	0.0202	—	[64]
syn	0.4347	0.0208	—	[64]
$\beta-C_{10}H_7CH$:				
anti	0.4711	0.0243	—	[64]
syn	0.4926	0.0209	—	[64]
	0.301	0.0132	—	[64]
$Ph-\ddot{C}-CH_3$	0.4957	0.0265	—	[62]
$Ph-\ddot{C}-CH_2Ph$	0.493	0.0289	—	[23]
$Ph-\ddot{C}-CO-Ph$	0.3815	0.0489	—	[23]
$Ph-\ddot{C}-Ph$	0.405	0.0194	~150°	[63, 66]
$CF_3-\ddot{C}-Ph$	0.5183	0.0313	—	[62]
$NC-CH$:	0.863	0.000	180°	[67, 68]
$(NC)_2C$:	1.002	0.002	180°	[69]
$HC\equiv C-CH$:	0.628	0.000	180°	[67]
$CH_3C\equiv C-CH$:	0.626	0.000	180°	[67]
$Ph-C\equiv C-CH$:	0.541	0.0035	—	[67]

Table 4 (continued)

Carbene	D/hc (cm^{-1})	E/hc (cm^{-1})	Angle	Ref.
$CH_3C{\equiv}C-C{\equiv}C-CH$:	0.609	0.000	180°	[67]
$(CH_3)_3C-C{\equiv}C-C{\equiv}C-CH$:	0.606	0.000	180°	[67]
$Ph-C{\equiv}C-C{\equiv}C-CH$:	0.533	0.000	180°	[67]
	0.4089	0.012	—	[69]
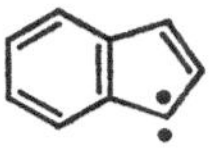	0.3777	0.0160	—	[69]
	0.4078 0.4084 0.4092	0.0283 0.0271 0.0283	135°	[65, 69] [70, 71] [71]
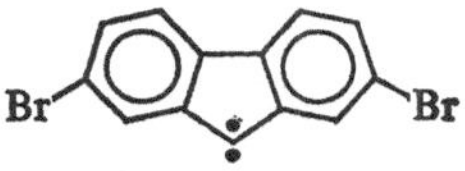	0.3991	0.0279	135°	[70, 71]
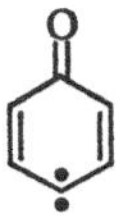	0.3179	0.0055	—	[72]
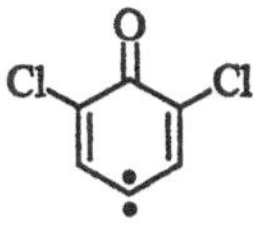	0.3284	0.0086	—	[72]
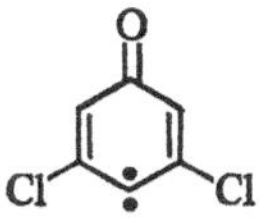	0.3470	0.0010	—	[72]
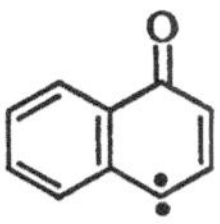	0.3333	0.0112	—	[72]
	0.40	0.02	—	[73]
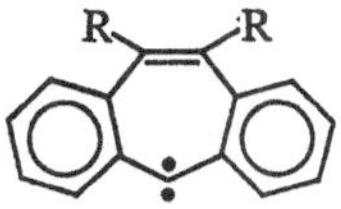 R = H	0.3787	0.0162	~150°	[71]
R = benzo	0.4216	0.0195	—	[71]
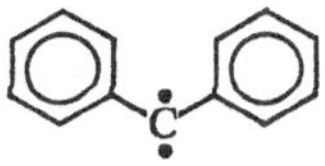	0.4050	0.019	~150°	[70, 71]

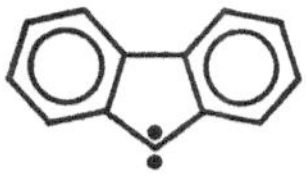

Table 4 (continued)

Carbene	D/hc (cm^{-1})	E/hc (cm^{-1})	Angle	Ref.
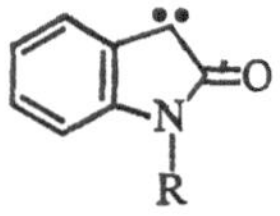 (indole-type carbene with =O, N–R)	0.38	—	—	[74]
 Ph–C̈–⟨C$_6$H$_4$⟩–C̈–Ph	0.0521	<0.002	—	[75]
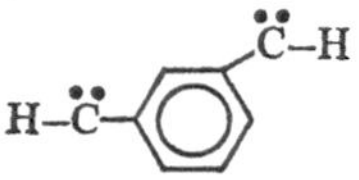 H–C̈–⟨C$_6$H$_4$⟩(C̈–H)	0.0844	0.0233	150°	[75]
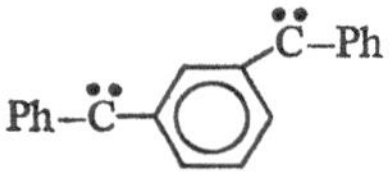 Ph–C̈–⟨C$_6$H$_4$⟩(C̈–Ph)	0.0701	0.002	150°	[75]

The ESR spectrum of methylene in a xenon matrix (4 °K) was studied by Wasserman [60]. Photolysis of diazomethane or diazirin gave an ESR-signal which persisted up to 20 °K but disappeared above 77 °K.

The values obtained were $D = 0.6844$ cm^{-1} and $E = 0.0034$ cm^{-1}. A species with a greater motional freedom was assigned to the values of $D = 0.6634$ cm^{-1} and $E \langle 0.002$. From these data an HCH angle of 136° was deduced, in excellent agreement with the latest calculations. However, the values of Skell [61], which are practically identical with the above, were interpreted differently. Similiar data obtained for HC̈D and DC̈D [60] are additional evidence that the bond angle of triplet methylene is in fact 136°.

ESR studies on alkyl-carbenes were carried out on fluoro-substituted carbenes only. The D-values in Table 4 are of the order of 0.7 cm^{-1} and the E-values range between 0.02 and 0.04 cm^{-1} [62]. The angle obtained for trifluoromethyl-carbene is $\approx 160°$, while for bis-(trifluoromethyl)-carbene it is $\approx 140°$, which again indicates that these carbenes are bent. The E-values of CF_3–C̈–Ph and CH_3–C̈–Ph differ appreciably, but the reason for it is not fully understood.

The arylcarbenes are also bent, the angle being about 150—155° for phenyl- and diphenylcarbenes [63,64]. The zero-field splitting parameters were shown to be appreciably dependent on the host matrix used [65]. However diphenylcarbene prepared from different precursors proved to be identical and is believed to be the triplet species of diphenylcarbene.

In arylcarbenes not only the internal bond angle but also the dihedral angle between the bond plane and the aryl plane can be bent. Photolysis of α- and β-naphthyl-diazomethanes in a matrix gave two isomeric carbenes, which produced different ESR signals [64].

ENDOR (electron nuclear double resonance) studies of diphenyl-carbene indicated a dihedral angle of 34° for this species [66].

$\Theta = 34°$
$\phi = 151°$

All carbenes mentioned so far were more or less bent. Carbenes containing triple bonds, such as cyano- or ethynylcarbenes had a zero E-value [67]. A similar result was obtained for dicyano-carbene [69]. The only possible interpretation of these results is a *linear structure* for these carbenes. The lower E-values for $Ph-C\equiv C-\ddot{C}H$ and $Ph-C\equiv C-C\equiv C-\ddot{C}H$ can be attributed to a partial delocalization of the electron in the p-orbital into the π-orbitals of the phenylgroup. The ESR-spectra of cycloalkene-carbenes [69-74] show that the ground states of carbena-cyclopentadienes, -cyclohexa-dienones and -cyclohepta-trienes are triplets, in conformity with theory [56]. The D-values are lowest for carbena-cyclohexadienones [72] indicating considerable electron transfer from the p-orbital to the carbonyl group. If the aromatic rings are annulated as in fluorenylidene, then delocalization is decreased. The reason for the relatively low E-values is believed to be one electron being mostly located in the σ-orbital, whereas the other p-electron is delocalized over the entire system. The E/D-ratio for cyclopentadienylidene, indenylidene and fluorenylidene [69-71] suggests an internuclear angle of about 135°. This value is, however, much larger than that of any angle which an undistorted cyclopentadienering may contain. It was accordingly concluded that the bonds in these carbenes are bent [69-71]. The bis-carbene obtained by photolysis of 1,4-bis(α-diazobenzyl)benzene gave zero-field splitting parameters (see Table 4) which indicated an about 4 Å separation of unpaired electrons.

2.2.2. Electronic Spectroscopy

Electronic spectra of carbenes were mostly obtained by flash photolysis in the gas phase, but some spectra were obtained in matrices at low temperatures.

The most interesting results have been obtained by Herzberg [31,76], on methylene itself but the original interpretation proved to be erroneous. A reinterpretation [77] — based on the ESR-data obtained by Wasserman [60] — has validated the results obtained by both groups. Flash photolysis of diazomethane under low pressure gave an absorption at 141,4 nm, which showed an additional fine structure. When the CH_2N_2/N_2 ratio was increased, a second photolytically generated species absorbing at 550—590 nm, appeared. The analysis of the rotational fine structure indicated that the first species was the *triplet methylene*. The band at 141,4 nm was due to a $^3\Sigma_g^- \rightarrow ^3\Sigma_u$ absorption; the structural parameters were: r_{C-H}-value $= 1.03$ Å; $H-\ddot{C}-H$-angle 136°. The band at 550—590 nm was assigned to a $^1A_1 \rightarrow ^1B_1$-transition of the *singlet methylene* having the following structural parameters: r_{C-H}-value $= 1.11$ Å; $H-\ddot{C}-H$-angle 102.4°. Matrix experiments have proved to be less successful [78].

Electron absorption and emission spectra of triplet diarylcarbenes have also been obtained. The emission of diphenylcarbene occurs around 480 nm. The ex-

101

citation fluorescence spectrum showed the same bands at 300 and 465 nm as the absorption spectrum. This indicates the presence of the same species. Since this spectrum was obtained in the matrix and disappeared on warming, it was assigned to the *triplet state* of diphenylcarbene [79]. Similar data for substituted diphenylcarbenes are shown in Table 5.

Table 5. Spectra of diphenylcarbenes (in 2-methyltetrahydro-furan) and cyclic carbenes

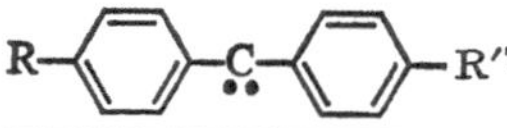 R—⟨⟩—C̈—⟨⟩—R′[79]		Absorption maxima (nm)	Emission maximum (nm)
R	R′		
H	H	300, 465	480
Cl	H	311, 475	487
Br	H	316, 475	488
CH$_3$	H	301, 472	487
OCH$_3$	H	335—345[1]	495
NO$_2$	H	265, 370, 555	no emission
Ph	H	355[1]	555
OCH$_3$	OCH$_3$	335—345[1]	507
 [80]		296	
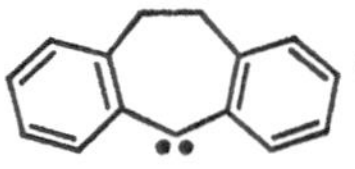 [81]		510	
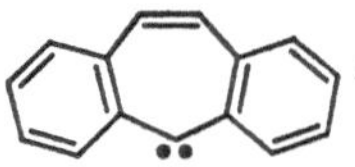 [82]		380 395 486 lifetime: 10 µsec (room temperature)	

[1] Obtained from the fluorescence excitation spectrum

Moritani [81,82] showed that the absorption spectra of dibenzo[a, d]cyclohep-tatrienylidene and 10, 11–dihydro dibenzo[a, d]cycloheptadienylidene can be recorded both at 77 °K and at room temperature. This proves that the same species is observed in the matrix and in solution, and is the triplet carbene. The lifetime of this carbene at room temperature was determined to be about 10 µsec.

The absorption maxima shown in Table 5 were also attributed to triplet carbenes.

2.2.3. Chemically Induced Nuclear Polarization (CIDNP)-Experiments

CIDNP proved to be a very promising new technique in studying the reactions given by triplet and singlet species. Whereas all other spectroscopic techniques merely provide information on the carbenes formed, CIDNP is the only technique for the investigation of the reaction mechanism of carbenes. Since the CIDNP method is relatively new the information collected so far is scanty. A simplified description of CIDNP is given below:

$$R-R \xrightarrow[\text{or } h\nu]{\Delta} R\cdot \ \cdot R \longrightarrow P_C$$
$$\longrightarrow P_E$$

P_C = cage or pair product
P_E = escape product

As a result of a sequence of chemical or photochemical reactions a pair of radicals is produced. This pair usually has the same spin as the precursor from which it originates, in accordance with the spin conservation law.

The radical pair thus formed — which is also termed "geminate pair" — now recombines to form the pair product P_C or it may escape from the solvent cage, form free radicals and form different products P_E by combination of these radicals. These free radicals have lost the original memory of the spin state of the precursor. On the other hand, the pair product generated in the solvent cage will remember the spin state of the radical pair, since both processes are fast in comparison with the time required for a change of the angular spin momentum. If the radical pair is a *singlet*, *much* of the pair product is produced whereas *little* pair product is obtained if the precursor is a *triplet* [83].

For a combination to occur, the radical pair must cross to a singlet state — by mixing of T and S states. Through the interaction of the odd electrons and the magnetically active nuclei, the nuclear spin states can influence the probability of intersystem crossing. Thus, a nuclear spin selection yields pair products, with a nonequilibrium distribution of nuclear spins. In other words, enhanced emission or absorption in NMR is given by the product (P_C) originating from the geminate radical pair.

Closs [83,84] developed a theory which quantitatively explains and predicts the CIDNP-effect. Kaptein [85] proposed qualitative rules to predict absorption or emission, *i.e.*, the phase of the emission (E) or absorption (A) signal. His equation is as follows:

$$\Gamma_{me} = \mu \cdot \Sigma \cdot A_i \cdot A_j \cdot \sigma_{ij} \begin{cases} +: E/A \\ -: A/E \end{cases}$$

μ $\begin{cases} + \text{ for triplet precursors and random fate recombinations} \\ - \text{ for singlet precursors} \end{cases}$

Σ $\begin{cases} + \text{ recombination} \\ - \text{ radical transfer products} \end{cases}$

103

$$\sigma_{ij} \begin{cases} + \ i \text{ and } j \text{ on the same radical} \\ - \ i \text{ and } j \text{ on different radicals} \end{cases}$$

A_i, A_j = hyperfine coupling constants of radicals A_i and A_j.

A positive sign means that an E/A signal will be produced whereas a negative sign indicates an A/E signal.

Reactions of triplet carbenes were first studied for diphenylcarbene, whose ground state is known to be triplet. Photolysis of diphenyl-diazomethane 1 in toluene yields the triplet diphenylcarbene 2. This carben can abstract a hydrogen atom from toluene to form the geminate radical pair 3. 3 is still in triplet state, which can undergo bonding only after intersystem crossing to the singlet state. The interaction with the nuclei then gives rise to an absorption spectrum followed by emission of the pair product 1,1,2-triphenylethane 5, $i.e.$, an A/E-phase spectrum is observed (see Fig. 5).

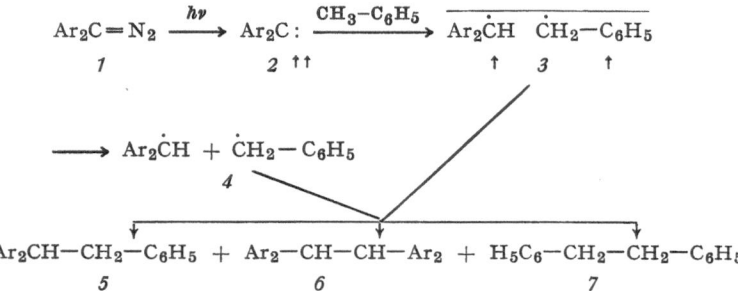

This can happen only if the precursor was a *triplet* radical pair. The 1,1,2,2-tetra-phenyl- (6) and 1,2-diphenyl-ethane 7 give no signals, clearly demonstrating that they are formed from free radicals by the escape route.

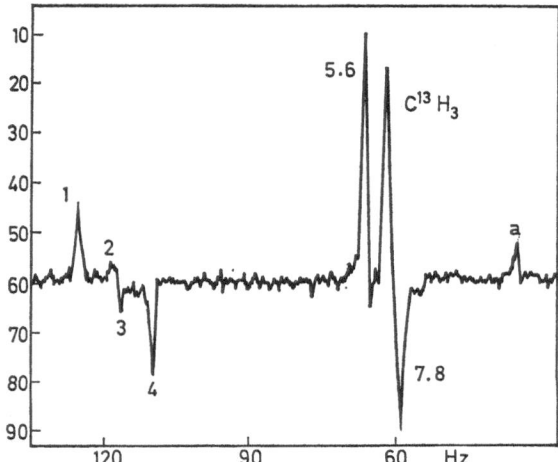

Fig. 5a. Spin-polarized spectra of 1: photolysis of diphenyldiazomethane in toluene. The chemical shift scale is in hertz downfield from the toluene methyl resonance [86]

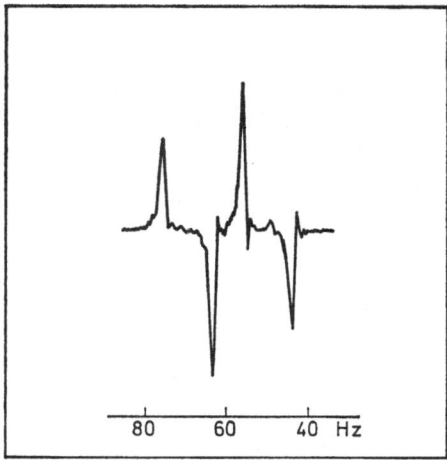

Fig. 5 b. Polarized spectrum obtained in reaction of *1* (0.2 M) in methyl phenylacetate at 140°. Chemical shifts are in hertz downfield from solvent CH_2 [86]

In the photolysis of *p,p'*-dibromo-diphenyl-diazomethane in toluene the geminate pair effect is observed [87]. However, it is accompanied by the enhanced emission by the escape product *7*. This means that the carriers of the original polarization were the free radicals, whose lifetime is obviously shorter than the nuclear relaxation time.

Photolysis of *p*-diazo-cyclohexadienones *8* in cyclohexane or carbon tetrachloride showed that the unsubstituted carbena-cyclohexadienone *9a* reacts as a singlet, whereas the 2,6-di-t-butyl-carbena-cyclohexadienone *9b* reacts as a triplet to give the products *12* and *13* [88].

Sensitizers have also been used to induce intersystem crossing to the triplet The insertion reaction of methylene into C—H bonds in solution is widely acknowledged to be a concerted process resulting from the singlet state. Photolysis of diazomethane in the presence of benzophenone, on the other hand, should result in an energy transfer with formation of a triplet methylene, which should react with C—H-bonds by an abstraction/recombination mechanism. A CIDNP study of this reaction in toluene again revealed a signal indicating the presence of methyl protons of ethylbenzene *14* in an AE-phase [89]. This leads to the conclusion that the precursor was in fact triplet methylene. The mechanism of this reaction can be represented as follows:

$$(H_5C_6)_2C{=}O \xrightarrow{\ h\nu\ } {}^1(H_5C_6)_2C{=}O$$

$${}^1(H_5C_6)_2C{=}O \xrightarrow{\ isc\ } {}^3(H_5C_6)_2C{=}O$$

$${}^3(H_5C_6)_2C{=}O + CH_2N_2 \longrightarrow {}^3CH_2N_2 + (H_5C_6)_2C{=}O$$

$${}^3CH_2N_2 \longrightarrow {}^3{:}CH_2 + N_2$$

$${}^3\ddot{C}H_2 + CH_3C_6H_5 \longrightarrow \overline{CH_3{\cdot}\quad {\cdot}CH_2C_6H_5} \longrightarrow CH_3{\cdot} + \dot{C}H_2{-}C_6H$$

$$CH_3{-}CH_2{-}C_6H_5 \qquad\qquad\qquad products$$

14

Photolysis of diazomethane in carbon-tetrachloride in the presence of benzophenone yields 1,1,1,2-tetrachloroethane showing an enhanced absorption due to the triplet carbene. The direct photolysis of diazomethane proceeds via singlet methylene [90]. CIDNP-studies of the photolysis of methyl-diazoacetate, for which a radical pair mechanism was suggested, were recently challenged [92].

3. Reactions of Triplet-Carbenes

Most of the reactions of triplet carbenes discussed in this chapter will deal with reactions in solution, but some reactions in the gas phase will also be included. Triplet carbenes may be expected to show a radical-like behaviour, since their reactions usually involve only one of their two electrons. In this, triplet carbenes differ from singlet carbenes, which resemble both carbenium ions (electron sextet) and carbanions (free electron pair). Radical like behaviour may, also be expected in the first excited singlet state S_1 (*e.g.* the ${}^1B_1$ state in CH_2) since here, too, two unpaired electrons are present in the reactive intermediate. These S_1-carbenes are magnetically inert, i.e., should not show ESR activity. Since in a number of studies ESR spectra could be taken of the triplet carbene, the reactions most probably involved the T_1-carbene state. However, this question should be studied in more detail.

3.1. Abstraction Reactions

3.1.1. Mechanism

Carbenes can formally insert in a various number of σ-bonds and this reaction is very general. It may be imagined to proceed according to three different mechanisms:

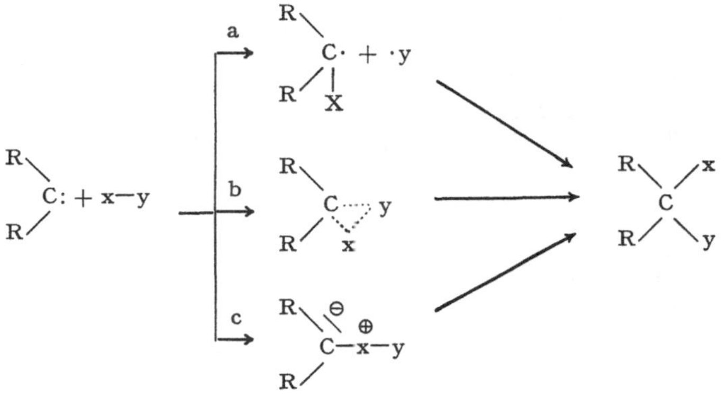

1. Abstraction of one atom of the covalent bond, with formation of two radicals, which then recombine (a),
2. concerted insertion via a three-membered transition state (b), and
3. formation of an ylide, which can rearrange to the final (formal) "insertion product".

Triplet carbenes react exclusively according to mechanism a), by abstraction-recombination involving single bonds such as C—H and C—Cl.

On the other hand, singlet carbenes react mostly by the concerted mechanism b) with C—H, C—C, C—X, N—H, O—H, S—H, M—C, M—X and M—M-bonds. Mechanism c) involving the ylide is less probable.

The mechanism proposed for carbene-abstraction and carbene-insertion reactions is based on the calculations of Dewar (MINDO/2) [52] and Hoffmann (extended Hückel) [93]. Hoffmann dealt only with the concerted reactions of singlet carbenes, whereas Dewar discussed both singlet and triplet carbene reactions. The calculations of Dewar [52] for the reaction of triplet methylene with methane gave the following results:

Two valleys for two different approaches were discovered corresponding to modes a) and b):

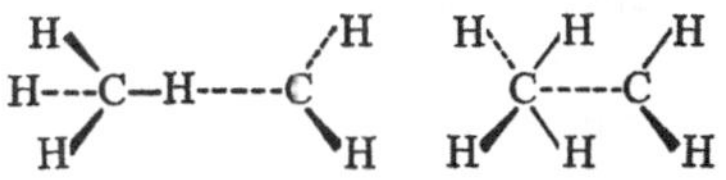

Mode a) should result in the abstraction of a hydrogen atom, and thus produce methyl radicals. In mode b) the energy steadily increases and no stable intermediate is formed. The abstraction reaction with formation of two methyl radicals

107

proceeds very readily, and requires only a small activation energy; the calculated value [52] is only 3.8 kcal/mole. No estimates of the heat of activation are as yet available (see Fig. 6).

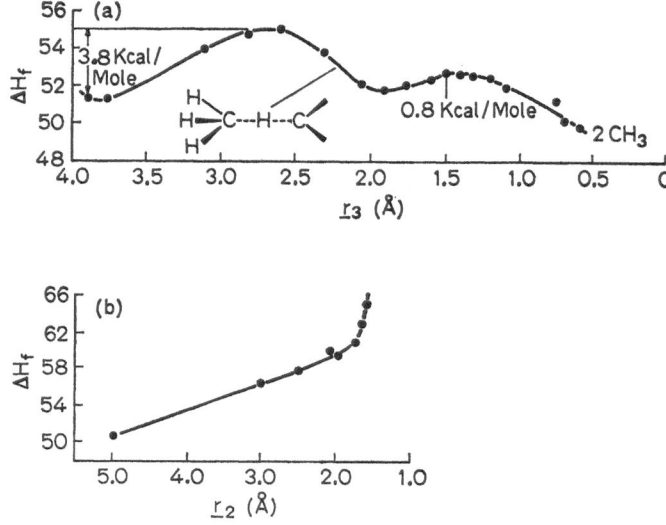

Fig. 6. (a) Reaction path for the process indicated in mode a); (b) corresponding path as in mode b)

3.1.2. Abstraction and Recombination Reactions

The insertion products obtained by the photolysis of diazomethane in the gas phase in the presence of alkanes also include products originating from ethyl radicals, the formation of which must be explained by postulating vibrationally excited species [94, 95]. The relative rates of abstraction of $^3CH_2$ from alkanes are $1°:2°:3°$ C—H-bonds $= 1:14:150$. The rate ratio of addition/$3°$—C—H-abstraction of $^3CH_2$ is $3.5/1$ [95].

However, collisional deactivation in solution is so effective that no vibrationally excited species is present. The reaction of photochemically generated methylene with 2-methylpropene-1-^{14}C yields, 2-methyl-butene, which is formed by allylic insertion. In the liquid phase 2% of the rearranged product labeled in the 3-position are formed, whereas in the gas phase 8% of this olefin can be isolated. This can be interpreted as follows: 4% of 2-methyl-butene in solution and 16% of 2-methyl-butene in the gas phase are formed by an abstraction-recombination mechanism involving triplet methylene [96].

Photolysis studies of diazomethane/isopentane mixtures in the presence and absence of oxygen, gave a calculated figure of $15—20\%$ triplet methylene [97].

Photochemically generated triplet methylene (from CH_2N_2) abstracts hydrogen from ethers almost exclusively at the α-position. The insertion/abstraction ratio was used to calculate the percentage of triplet methylene; it was about 10%.

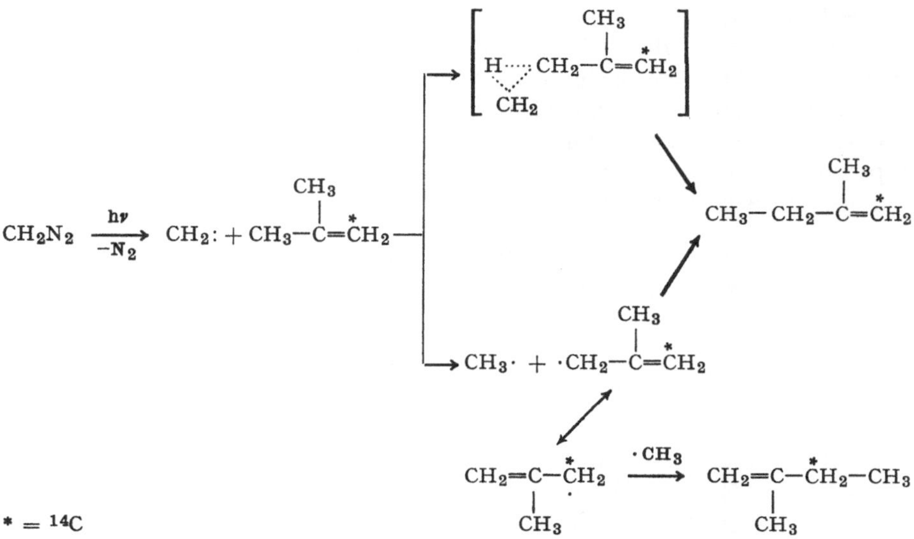

$$\left[\begin{array}{c} CH_3 \\ | \\ H\cdots CH_2-C=\overset{*}{C}H_2 \\ | \\ CH_2 \end{array} \right]$$

$$CH_2N_2 \xrightarrow[-N_2]{h\nu} CH_2: + CH_3-\overset{CH_3}{\underset{|}{C}}=\overset{*}{C}H_2 \longrightarrow$$

$$CH_3-CH_2-\overset{CH_3}{\underset{|}{C}}=\overset{*}{C}H_2$$

$$\longrightarrow CH_3\cdot + \cdot CH_2-\overset{CH_3}{\underset{|}{C}}=\overset{*}{C}H_2$$

$$CH_2=\overset{*}{C}-\overset{\cdot}{C}H_2 \xrightarrow{\cdot CH_3} CH_2=\overset{*}{C}-\overset{\cdot}{C}H_2-CH_3$$
$$\underset{CH_3}{|} \qquad\qquad \underset{CH_3}{|}$$

* = ^{14}C

Fluorinated hydrocarbons favoring intersystem crossing by dilution increase the proportion of triplet methylene up to about 56% [98].

Evident cases of abstraction/recombination mechanism are observed with phenylsubstituted carbenes. Diphenyl-diazomethane, which is photolyzed to give the triplet diphenyl-carbene, very readily abstracts a hydrogen atom from the benzyl group of toluene. The primarily formed radicals can now recombine to give a formal "insertion product" — 1,1,2-triphenylethane — or they can recombine to form 1,1,2,2-tetraphenylethane and 1,2-diphenylethane [99, 100].

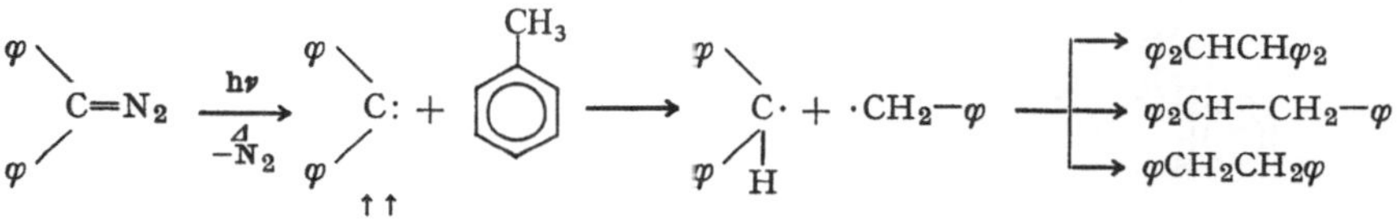

$$\overset{\varphi}{\underset{\varphi}{>}}C=N_2 \xrightarrow[-N_2]{h\nu} \overset{\varphi}{\underset{\varphi}{>}}C: + \text{(toluene)} \longrightarrow \overset{\varphi}{\underset{\varphi}{>}}\overset{|}{\underset{H}{C}}\cdot + \cdot CH_2-\varphi \longrightarrow \begin{cases} \varphi_2CHCH\varphi_2 \\ \varphi_2CH-CH_2-\varphi \\ \varphi CH_2CH_2\varphi \end{cases}$$

Flash photolysis of 5-diazo-10, 11-dihydro-dibenzo[a, d]cycloheptene (*15*) — which can be regarded as a bridged diphenyl-carbene — at room temperature in liquid paraffin first produced the spectrum of the triplet carbene *16*, which then disappeared to give the electronic spectrum of the radical *17*. The latter finally gave the dimer 5,5′-bi (10, 11-dihydrodibenzo[a, d]cycloheptenyl) *18* [82].

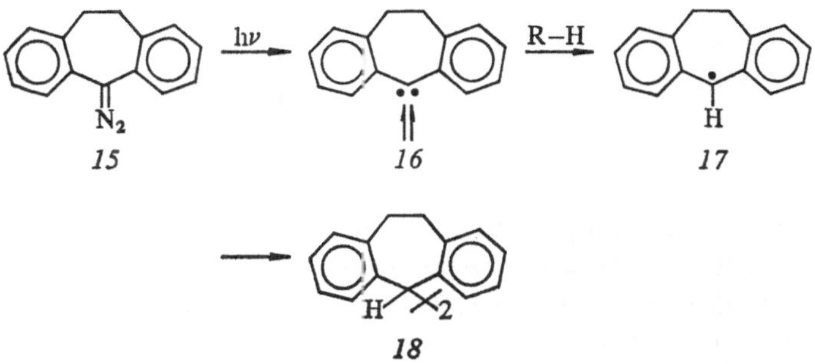

15 *16* *17*

18

109

The triplet carbenes fluorenylidene *20* and anthronylidene *24*, which can be generated from the diazoalkanes by photolysis, show a similar behaviour. Fluorenylidene in cyclohexane yields 9-cyclohexenyl-fluorene *21*, and 9,9'-difluo-renyl *22* which are clearly formed by an abstraction-recombination process [101]. Another example is anthronylidene *24* in cyclohexane or toluene, which yields the products *25, 26, 27* resulting from an abstraction-recombination process [102, 103]. Benzene, on the contrary, failed to give the radical pair product *20* [102].

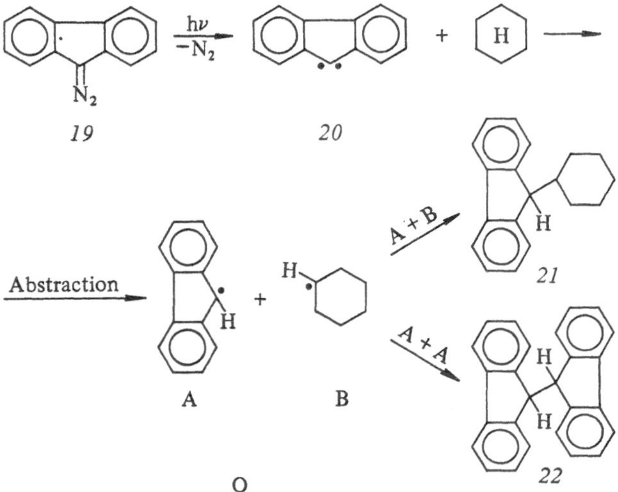

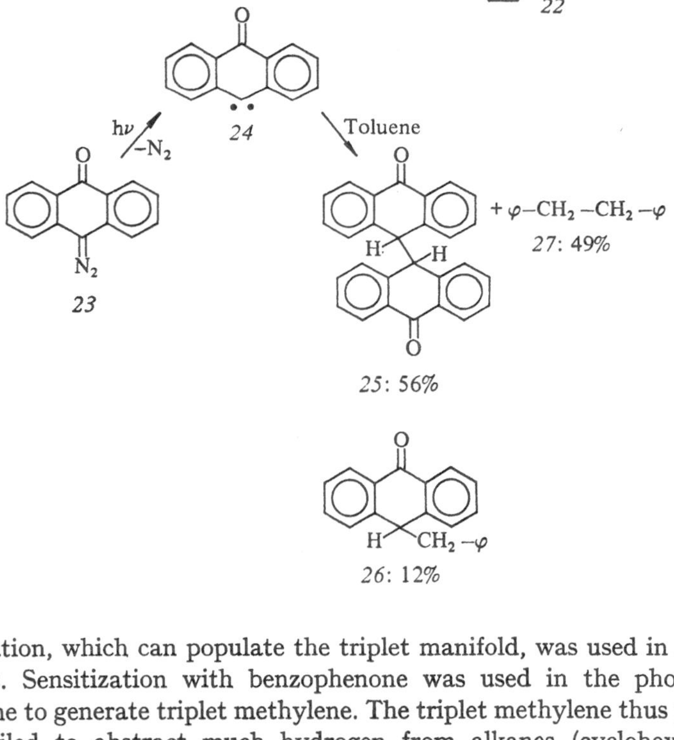

Sensitization, which can populate the triplet manifold, was used in a number of instances. Sensitization with benzophenone was used in the photolysis of diazomethane to generate triplet methylene. The triplet methylene thus produced, however, failed to abstract much hydrogen from alkanes (cyclohexene), but underwent polymerisation to polymethylene instead [104]. Sensitization was used to generate bis(methoxycarbonyl)carbene. The benzophenone-sensitized photolysis of dimethyldiazomalonate in 2,3-dimethylbutane gave a lower yield of product as in the direct reaction [105]; diethylmalonate and 1,1,2,2-tetracarbomethoxy-ethane were obtained in addition to direct photolysis products [105]. The reported data indicate that the triplet produces respectively, 23% and 38% of the abstraction products 30 and 31, which are absent in the singlet reaction. The selectivity as measured by the 3°/1°-insertion ratio increased from 13 to 20 in the triplet run (see Table 6). The singlet reacts with dimethyl ether via the ylid mechanism c), whereas the triplet carbene does not give this reaction (see also Ref. [106]). Abstraction products such as 29 are also formed in the photolysis of 28 in cyclohexane in the presence of thiobenzophenone [107].

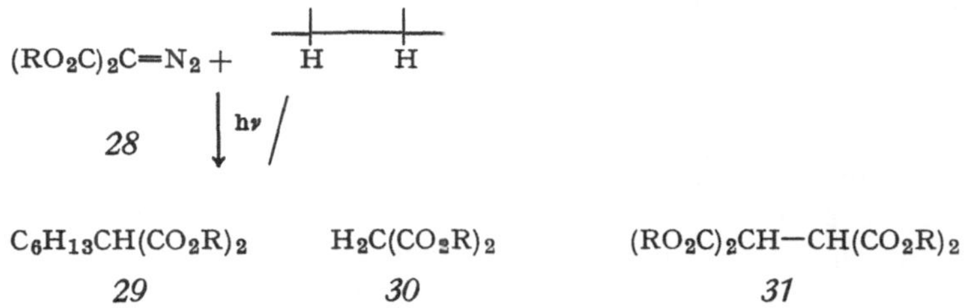

$$(RO_2C)_2C=N_2 + \quad \overset{|}{\underset{H}{}} \quad \overset{|}{\underset{H}{}}$$

$$28 \quad \Big\downarrow{}^{h\nu} \Big/$$

$$C_6H_{13}CH(CO_2R)_2 \qquad H_2C(CO_2R)_2 \qquad (RO_2C)_2CH-CH(CO_2R)_2$$

$$29 \qquad\qquad\qquad 30 \qquad\qquad\qquad\qquad 31$$

Table 6. Direct and sensitized photolysis of 28

		29	30	31
Direct		46%	trace	trace
Sensitized (Ph$_2$CO)		13%	23%	38%
Relative "3°/1°-insertion"	dir.	13		
Rate	sens.	20		

Whereas singlet carbenes insert in C—H-bonds with retention, no such studies have been carried out on triplet carbenes. Atomic carbon can also react in one of its three electronic states, vic. the triplet ^3P and the singlet ^1D and ^1S. Insertion of monoatomic carbon in C—H-bonds initially gives a carbene, whose subsequent fate depends on its multiplicity. ^3P (ground state) carbon atoms first form a triplet carbene, which then abstracts hydrogen from isobutane to finally form a methyl group in the molecule concerned, i.e. in isobutane. This is, however, the minor reaction (5% yield). Singlet ^1S carbon gives dimethyl-cyclopropane and 4-methyl-1-butene [109].

$$\overset{\diagdown}{\diagup}C-H \xrightarrow[(3P)]{\uparrow\downarrow C\uparrow\uparrow} \overset{\diagdown}{\diagup}\overset{\overset{H}{|}}{C}-\overset{\uparrow\uparrow}{C}: \xrightarrow{\overset{\nearrow}{=}\mathbf{H}} \overset{\diagdown}{\diagup}-CH_2\cdot \xrightarrow{\overset{\nearrow}{=}\mathbf{H}} \overset{\diagdown}{\diagup}-CH_3$$

$$\xrightarrow[(1S)]{\uparrow\downarrow C\uparrow\downarrow}$$

$$5\%$$

$$\triangleright\!\!\!\triangleleft \ \ + \ (CH_3)_2CH-CH=CH_2$$

$$52\%$$

In contrast to singlet carbon ^3P carbon does not add to double bonds [108].

3.2. Addition Reactions of Triplet Carbenes

3.2.1. Mechanism

Carbenes can add to *cis*- or *trans*-olefins to form cyclopropanes in two different ways: the original stereochemistry of the olefin may be retained, when the reaction is stereospecific, or the original stereochemistry of the olefin may be lost, in which case the reaction is regarded as nonstereospecific. This fact was recognized at a very early date by Skell [110], who postulated the following rules for $[1+2]$-cyclo-addition of a carbene to an olefin:

1. Singlet carbenes add stereospecifically to olefins in a concerted reaction mechanism.
2. Triplet carbenes add in a two-step reaction in a nonstereospecific manner.
3. Triplet carbenes react with 1,3-dienes more rapidly than with monoolefins.

Skell's hypothesis proved to be extremely useful in carbene chemistry even though it was frequently opposed. The principal significance of these rules is represented in the scheme below. The *singlet* reaction occurs in a *concerted* step, the cis-addition product being formed in a *stereospecific* manner. In the *triplet* addition, which is a two-step reaction, rotation is thought to be faster than inter-system crossing (spin inversion) and ring closure, *i.e.*, $k_3 > k_1 \sim k_2$, which would result in a mixture of *cis*- and *trans*-cyclopropanes. The reaction is accordingly *non-stereospecific*.

If carbenes with two different R groups are used, syn- and anti-cyclopropanes are obtained. In most cases cyclopropane, which is thermodynamically less stable, is predominant. Such reactions are called *stereoselective*.

These simple rules were recently shown to have a theoretical foundation. Hoffmann [109], who used EH- and Dewar [52] who used MINDO/2-calculations, laid down a theoretical base for the Skell hypothesis.

Hoffmann was the first to apply the concept of orbital symmetry to the cyclo-addition of carbenes to olefins. This concept, which is based on EH-calculations was demonstrated for the $[1+2]$-cyclo-addition of triplet methylene to ethylene.

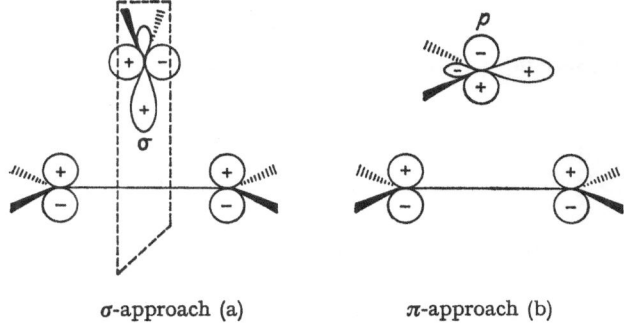

R\(_\), R'\(_\): C:‖ →[Concerted addition] R' R' / R' R'

Singlet

Singlet: *cis*—addition stereospecific (a)

Triplet

1. Intersystem crossing k_1
2. Ring closure

Rotation k_3

cis—Addition ⎬ (b) Nonstereo—specific

1. Intersystem crossing k_2
2. Ring closure

trans—Addition

Two different approaches are possible: through a σ-orbital of methylene (a); through a p-orbital of methylene (b).

The σ-approach is the more symmetrical, since it has C_{2v}-symmetry.

σ-approach (a)   π-approach (b)

Fig. 7. σ- and π-approach of a carbene to an olefin

The plane indicated serves to construct the orbital correlation diagram using the methylene ground state $^3B_1$ or $\sigma\,p$-configuration, the lowest singlet being $^1A_1$ or σ^2 (see Fig. 8).

From the state correlation diagram based on Fig. 9 it can be seen that the triplet state $^3B_1$ of methylene and the ground state of ethylene correlate with

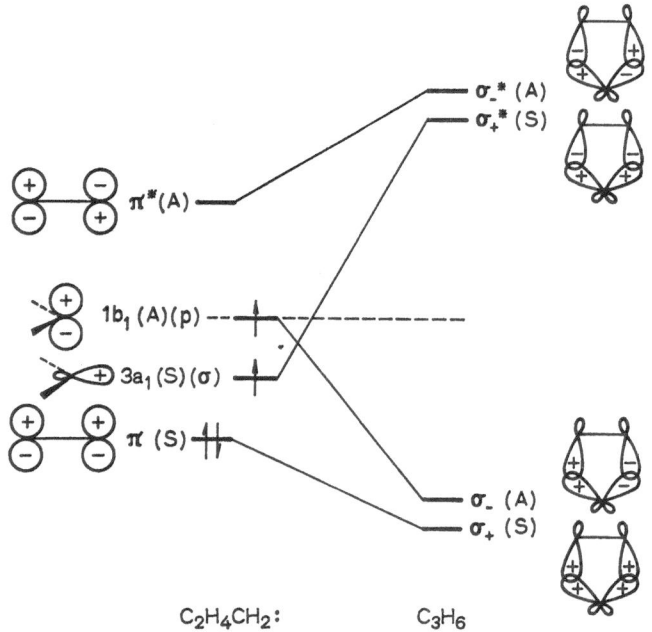

Fig. 8. Orbital correlation diagram for the addition of methylene to ethylene through transition state a (Fig. 7) [109]

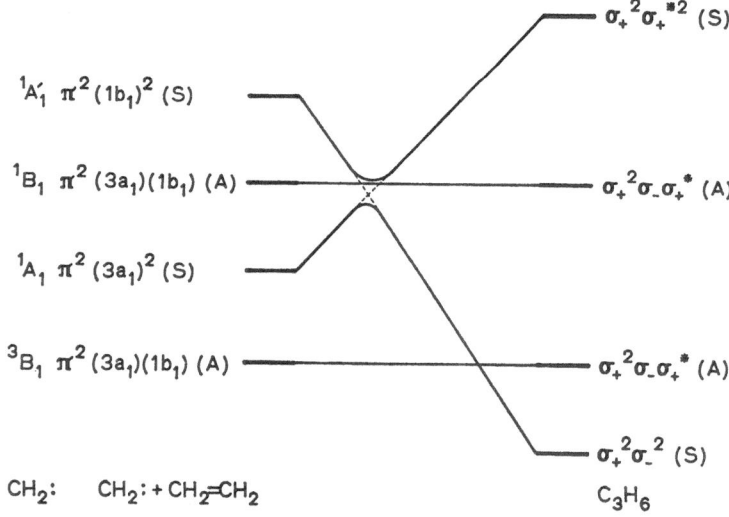

Fig. 9. Schematic state correlation diagram for the addition of methylene to ethylene. Bracketed letters indicate the total symmetry (S) or antisymmetry (A) of each state

the first excited state of cyclopropane which, according to the EH-calculations, is in fact the trimethylene diradical. Here no barriers to rotation exist and the stereochemistry is lost in this reaction. The *singlet* methylene $^1A_1$ and the ground state of ethylene, on the other hand, correlate only with the ground state σ_+^2 σ_-^2-configuration of cyclopropane, because of the non-crossing rule. This reaction therefore requires a large activation energy, *i. e.*, is a prohibited reaction. However these results hold only for the most *symmetrical σ-approach* of methylene to the olefin. Hoffmann [109] and Dewar [52] both calculated the different phases of the geometry of the methylene approach to olefins. The successive positions of the triplet methylene approach are shown in Fig. 10.

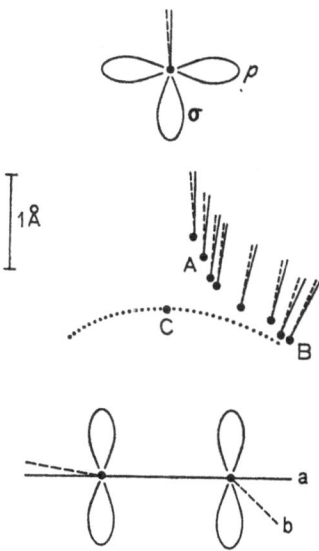

Fig. 10. Geometry of approach of T_1 carbene to ethylene. Lines a and b indicate the initial and final position of the hydrogen atoms at the central carbon atom of the resulting biradical $\cdot CH_2CH_2CH_2 \cdot$. C being the transition state [52]

The methylene triplet adds to ethylene symmetrically through the σ-approach (A). Then, at a closer distance, it bends to one side, forming the triplet biradical directly (B). The central methylene group (formed from ethylene) is bent downwards by this process (Fig. 10). At this stage rotation or direct ring closure can occur, with loss of stereochemistry following bond formation to yield cyclopropane. The cyclo-addition of the triplet requires only a small activation energy of about 5 kcal/mole [52].

The singlet carbene, on the other hand, can add through a less symmetrical transition state in a slightly bent *p*-approach, thus circumventing the barrier imposed by orbital symmetry:

These results can be summarized as follows: The *triplet carbene* ($^3B_1$) adds *nonstereospecifically* because its complex and a ground state ethylene correlate with the triplet state of an excited trimethylene configuration, which has no barriers to rotation around terminal bonds.

115

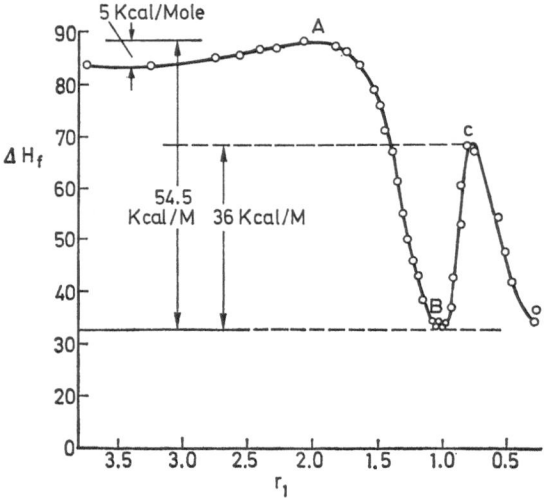

Fig. 11. Plot of the calculated heat of formation (H_1; kcal/mole at 25°) vs. the reaction coordinate r_1 for addition of triplet (T_1) carbene to ethylene and for rearrangement of the resulting biradical [52]

The *singlet carbene* ($^1A_1$) adds *stereospecifically*, because it can correlate with the lowest singlet configuration of a trimethylene diradical, and thus also with a ground state of cyclopropane [109].

For the first excited singlet S_1 ($^1B_1$) a nonstereospecific addition is also predicted.

3.2.2. Stereospecifity of Carbene [1+2]-Cycloaddition

The experimental evidence most often employed to decide between the singlet and triplet multiplicity of a carbene is [1+2] cycloaddition to cis- or trans-olefins. The olefins employed in these studies are mostly 2-butenes or 4-methyl-2-pentenes, but in a few cases other olefins were also utilized.

Gas Phase Reactions

Photolysis of diazomethane in the gas phase produces $:CH_2$ which can add to cis-2-butene to form dimethyl-cyclopropanes *32* and *33*. The insertion products *34* and *35* are also formed [110,111].

$$CH_2N_2 \xrightarrow{h\nu/gas\ phase} :CH_2$$

32   *33*

34   *35*

Since the addition of methylene to an olefin should be exothermic, with the evolution of about 90 kcal/mole, isomerisations of the initially formed cyclopropanes are very likely, since they only need about 64 kcal/mole. RRKM-studies demonstrate that this isomerisation should be faster than the rearrangement of cyclopropanes 32, 33 to the pentenes 34, 35 [112]. Numerous studies of the photochemical generation in the gas phase provided conclusive evidence in favour of these findings [110,111,113].

If the photolysis of diazomethane and cis-2-butene is carried out under a low pressure in an inert gas, the addition products 32 and 33 as well as the insertion products 34, 35 are formed. Increasing the pressure of the inert gas (Ar, Xe, N_2 etc.) reduces the yield of insertion product to zero and the yield of the trans-cyclopropane 33 also decreases. The reaction is more stereospecific. However, a further increase in the amount of the inert gas again increases the amount of trans-cyclopropane 33, i.e., the reaction becomes less stereospecific. These results are explained by the primary formation of vibrationally excited cyclopropanes with enough energy to isomerize to the other geometrical isomer and to the pentenes 34, 35. At medium values of the inert gas pressure complete thermal deactivation of cyclopropanes 32 and 33 occurs, and only the addition products can be isolated. At higher pressures the intersystem crossing of singlet to triplet methylene is favored, thus increasing the "non-stereospecifity" of the reaction. This explanation is supported by the flash spectroscopic experiments of Herzberg who observed a similar increase in the concentration of triplet $\check{C}H_2$ on the addition of inert gas [76,77]. The stereospecifity of the cyclo-addition of methylene to cis- or trans-olefins can be affected by the presence of oxygen. O_2 serves as a triplet trap and the stereospecifity — now due only to singlet methylene — increases [115]. The addition of methylene to cyclobutene was used to estimate the percentage of triplet present in the gas phase; it was about 20—30% [116]. Results of ketene photolysis yielded a similar value [117].

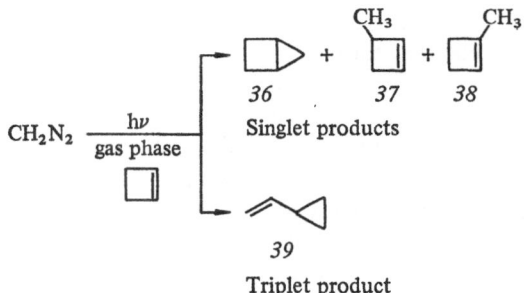

36   37   38
Singlet products

39
Triplet product

Addition Reactions in Liquid Phase

Photolysis of diazoalkanes in liquid phase yields carbenes in a vibrationally relaxed state, since deactivation in solution immediately removes all excess vibrational energy. The addition of carbenes to the olefins, which results in non-stereospecific formation of cyclopropanes, must therefore result from the different multiplicity of carbenes — singlet or triplet. Since most of these multiplicity studies are photochemical, the following scheme again shows the formation of cyclopropanes from the photochemically generated carbene.

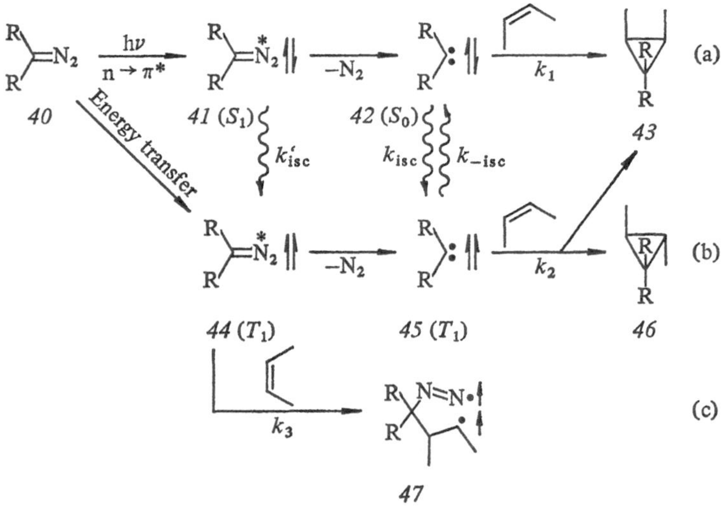

The experimental ratio of cis- to trans-cyclopropane $43:46$, $i.e.$ the stereospecifity of the reaction cannot be considered as a simple indication of singlet or triplet percentage of $R_2C:$, since the stereochemistry of the cyclo-addition depends on many factors. Photolysis produces the excited S_1-state of the diazoalkane 41. This compound can lose nitrogen and form the singlet carbene 42 (S_0-state). 42 can add directly in a stereospecific manner if k_1 is large. If, however, intersystem crossing $42 \rightsquigarrow 45$ (k_{isc} is large) competes favorably with $42 \xrightarrow{k_1} 43$, then the triplet state of carbene 45 is populated, and the carbene can now add non-stereospecifically to the olefin to give 43 and 46.

This simple route can, however, become more complicated if intersystem crossing occurs already in the excited diazoalkane, $e.g.$ $41 \rightsquigarrow 44$ (k'_{isc} is large). This means that the triplet carbene 45 is generated by two different routes. It is thus not possible to deduce the relative amount of singlet and triplet carbenes from the ratio of cis- and trans-cyclopropanes 43, 46 even if equal reaction rates k_1 and k_2 for the cyclo-addition are assumed, which is certainly not always the case. On the other hand, the triplet state of a carbene is populated only by the intersystem crossing steps k'_{isc} and k_{isc}. It depends on the effectiveness of these steps if a triplet carbene is produced to a larger or a smaller extent. We may summarize as follows. Different rates of cycloaddition to the olefin must be assumed for singlet and triplet carbenes. $[1+2]$-Cycloaddition of triplet carbenes to olefins is always accompanied by abstraction reactions. Carbenes with a triplet ground state can nevertheless react as singlet carbenes, $i.e.$, stereospecifically, if they are generated at room temperature and if k_1 is larger than k'_{isc} and k_{isc}.

Table 7 is a synopsis of cycloaddition reactions of various triplet carbenes to cis- or trans-olefins which result in nonstereospecific additions.

Table 7. Reactions of triplet-carbenes in solution

Mode of Carbene generation	Stereospecifity		Ref.
	% trans-Cyclo-propane from cis-olefin	% cis-Cyclo-propane from trans-olefin	
CF_3CHN_2, hν	4 [1]	2 [1]	118)
HC≡C—CHN_2, hν	13 [1]	5 [1]	119)
$C_6H_5CHN_2$, hν	3 [1]	4 [1]	120, 121)
$(C_6H_5)_2CN_2$, hν	13 [1]	little	120, 121)
	35 [7]	—	121)
$(CH_3)_3SiCN_2CO_2C_2H_5$, hν	5 [1]	—	122)
$N_2C(CO_2CH_3)_2$, hν	8 [1]	10 [1]	105, 123)
$N_2C(CN)_2$, Δ (80 °C)	70	30	124)
Δ Cyclohexan	70 [1]	30 [1]	124)
$C_6H_5COCHN_2$, hν	50 [1]	27 [1]	125)
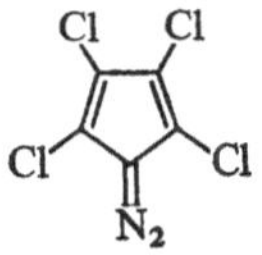 (tetrachlorocyclopentadienylidene, N₂, hν)	10 [1] 22 [3] 58 [4]	Trace [1] — —	126)
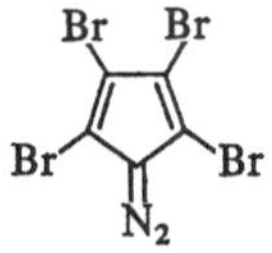 (tetrabromocyclopentadienylidene, N₂, hν)	55 [2]	0 [2]	127)
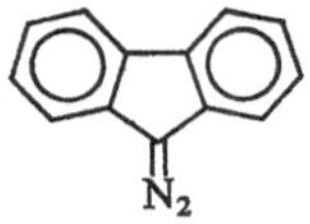 (fluorenylidene, N₂, hν)	34 [1] 45 [2] 77 [1]	0 [1] Trace 12 [1]	128)
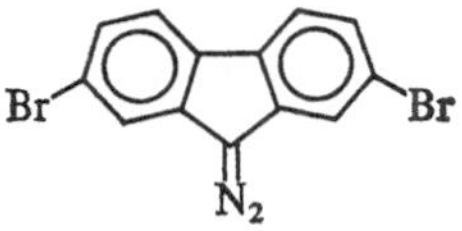 (dibromofluorenylidene, N₂, hν)	54 [4]	—	70)
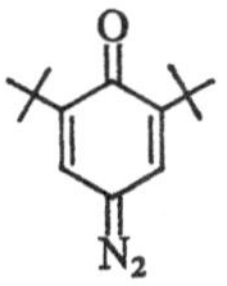 (di-tert-butyl cyclohexadienone diazo, N₂, hν)	5 [1]	3 [1]	129)

Table 7 (continued)

Mode of carbene generation	Stereospecifity		Ref.
	% trans-Cyclopropane from cis-olefin	% cis-Cyclopropane from trans-olefin	
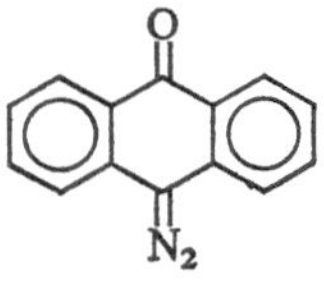 (structure) hν	Not clear: photochemical cis-, trans-isomerization not excluded		102, 130, 132)
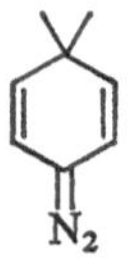 (structure) hν	8 2)	1 2)	73)

1) cis- or trans-But-2-ene.
2) cis- or trans-4-Methyl-pent-2-ene.
3) cis- or trans-Pent-2-ene.
4) Dialkyl maleate or fumarate.
5) cis-β-Deuteriostyrene.

It is seen from Table 7 that all these carbenes in solution react *nonstereospecifically*, indicating *triplet* character. For some carbenes such as $CF_3\ddot{C}H$, $Ph\ddot{C}H$, $(CH_3)_3Si\ddot{C}(CO_2C_2H_5)$, 2,6-di-t-butyl-carbena-cyclohexadienone and 4,4-dimethyl-carbena-cyclohexadienone, the degree of non-stereospecifity is quite small. Other carbenes usually contain π-bonds, aromatic rings or heavy atoms as functional groups.

These groups favour intersystem crossing, producing varying amounts of triplet carbene. The reasons for the heavy atom effect favouring k_{isc} are well known to be due to an increased spin-orbit coupling. Annulated aromatic rings lower the energy of the LUMO, bringing about increased interaction with the p_x-orbital of the carbene. The smaller energy difference between σ- and p-orbitals favours the triplet carbene. Photolysis of a diazoalkane at a low temperature should yield the triplet carbene. Moss et al. [133] generated triplet phenylcarbene by photolysis of phenyl-diazomethane in cis-butene matrices at a low temperature. The addition of the phenylcarbene thus produced was less stereospecific; the abstraction/addition ratio also increased. Triplet phenylcarbene is thought to be responsible for this behaviour.

The addition of the diphenylcarbene, generated by the photolysis of diazo-diphenylcarbene, to cis- or trans-butene gives mostly the abstraction product, and a small amount of the addition product. However, the abstraction processes are

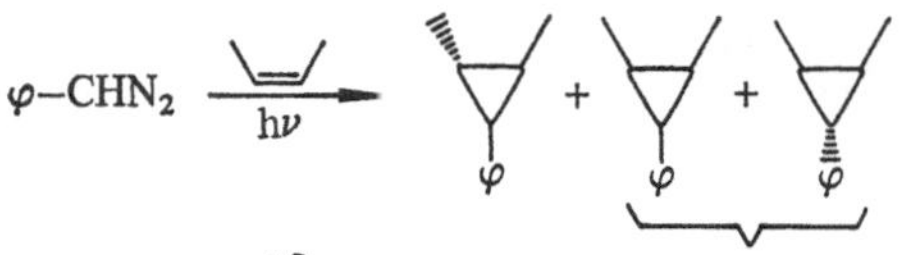

$\varphi-CHN_2 \xrightarrow[h\nu]{}$ [structures]

	0°	1.8–2.5	91.3–93.8
	−192°	2.0–4.3	40.0–47.6

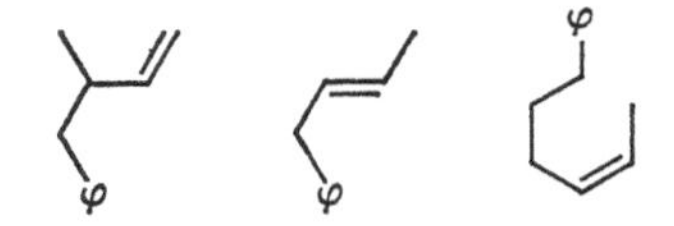

	0.4–2.6	0.3–0.4	2.3–4.4
	13.7–22.4	3.3–6.7	24.8–32.1

suppressed if styrene is used. Thus, the cis-β-deuterostyrene gives 65% of cis-cyclopropane *48* and 35% of trans-cyclopropane *49* [121].

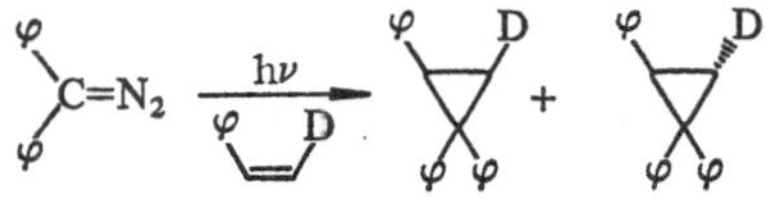

[structure] $C=N_2 \xrightarrow[h\nu]{}$ [structures]

48: 65%   *49*: 35%

If identical addition rates for singlet and triplet diphenylcarbene are assumed, this ratio corresponds to 70% of the triplet species.

Anthronylidene *64* forms no cycloaddition products with cis- or trans-4-methyl-2-pentene, but [1+2]-cycloaddition is observed with stilbene derivatives. This is also attributed to the reacting triplet state [102,130–132].

3.2.3. Dilution Experiments Resulting in Intersystem Crossing to Triplet Carbenes

The dilution technique makes use of the different concentration dependence of a) S-T-intersystem crossing, and b) [1+2]-cyclo-addition of a carbene to an olefin. The decay of the metastable singlet state is monomolecular, while the stereospecific addition is of the first order with respect to the concentration [26]. At high dilution with an inert solvent such as hexafluorobenzene or octafluoro-cyclobutane etc., the same cis-/trans-cyclopropane ratio should be obtained with cis- or trans-olefin as the starting compound.

This technique has been repeatedly utilized. Photolysis of diazomethane in the liquid phase yields methylene in the singlet state, as is shown by the practically stereospecific addition to cis- or trans-butene. Dilution with perfluoropropane reduces the degree of stereospecificity as well as the amount of C—H-insertion, indicating that triplet methylene is involved [134]. A similar effect has been reported for $CF_3\ddot{C}H$ which, on dilution with perfluoro-diethylether (in the gas phase), adds in a non-stereospecific manner owing to the presence of the triplet [118].

121

Table 8. Liquid phase photolysis of diazomethane in the presence of cis-2-butene [134]

Products	cis-2-Butene neat	cis-2-Butene + C$_3$F$_8$ (1:200)
cis-1.2-Dimethylcyclopropane	47.5	60.4
trans-1.2-Dimethylcyclopropane	0.4	13.3
cis-2-Pentene	39.1	9.3
trans-2-Pentene	0.0	7.1
2-Methyl-2-butene	12.5	1.9
2-Methyl-1-butene	0.3	1.9
3-Methyl-1-butene	0.2	6.1

In the photolysis of diphenyl-diazo-methane in olefins the ratio cyclopropane/abstraction products depends on the structure of the olefin. The nonstereospecific addition of diphenylcarbene to cis-2-butene is temperature-dependent [18b].

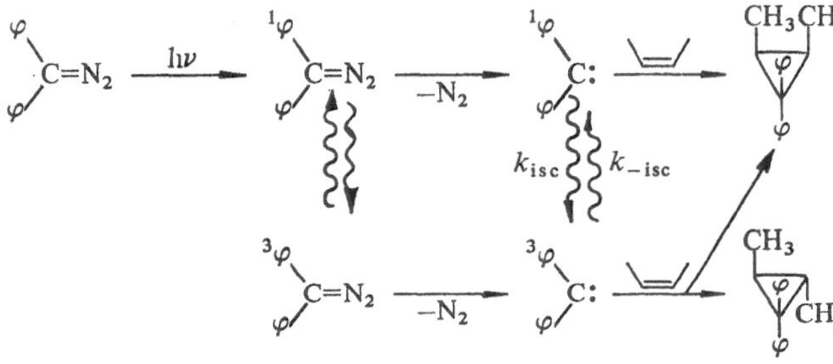

temp : ratio *57/58*

0°C   3,2

−66°C   9,0

At a given temperature, the cis-*57*/trans-*58* ratio is independent of the olefin concentration over a range of 150-fold dilution with cyclohexane. Hexafluorobenzene as diluent results in a more stereospecific reaction [18b].

The photolysis of diphenyl-diazomethane in cis-β-deutero-styrene, on the other hand, can be affected by dilution with hexafluorobenzene. The amount of trans-cyclopropane *49* is slightly larger, indicating that about 12% of the singlet carbene are still present in very dilute solutions (see Table 9).

This experimental fact indicates that the singlet and triplet diphenylcarbene are in thermal equilibrium and that k_{isc} is fast because the ^{3}Ph$_2$C̈-state is the ground state. The relatively low nonstereospecifity can be explained by assuming a) a faster addition of the singlet or b) most of the triplet undergoing the abstraction reaction and only little of it the addition reaction.

If, during the photolysis of methyl-diazo-malonate in cis-4-methyl-2-pentene, increasing concentrations of hexafluorobenzene are added, only very large amounts of solvent will affect the ratio of the cis-trans-isomers *50* and *51* [135].

Table 9. Stereochemistry of addition of diphenylcarbene to styrene as a function of added hexa-
flurobenzene

Mol % of Hexaflurobenzene	% cis Adduct	% trans Adduct
0	65	35
36	60	40
50	59	41
80	56	44
90	55	45
95	56	44

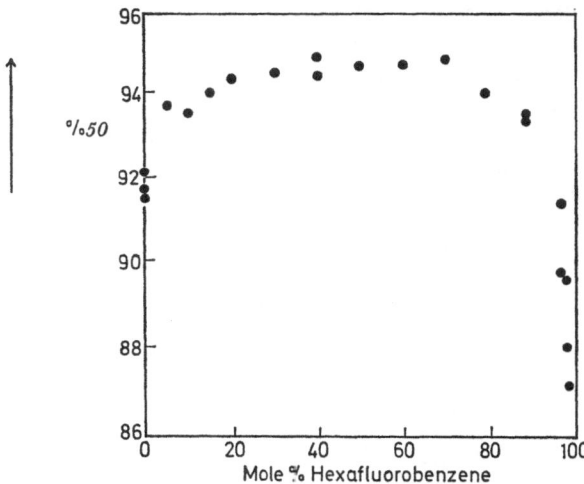

$(CH_3O_2C)_2C=N_2$ ⟶

$H_3CO_2C \quad CO_2CH_3$ $H_3CO_2C \quad CO_2CH_3$

28   50   51

Initially, the amount of the cis-isomer 50 increases slightly at lower concentrations of C_6F_6. An excited methyl-diazo-malonate — probably in the singlet
state — is thought to be responsible for the increased stereospecifity (see Fig. 12).

Fig. 12. Addition of 28 to cis-4-methyl-2-pentene in hexafluorbenzene

The yield of trans-cyclopropane 54 produced by addition of carbena-cyclohexadienone 52' to cis-2-butene increases from 5 to 33% if 90% C_6F_6 is added [136].

Fluorenylidene — which was photochemically generated from diazofluorene —
indicated a larger content of triplet carbene [128] (see Table 7) when diluted.

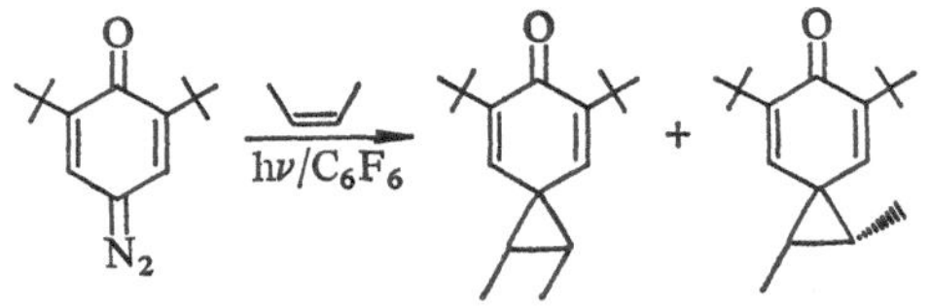

	52	C₆F₆ :	53	54

C_6F_6 :

	53	54
0%	95	5
90%	66	33

No *S-T*-intersystem crossing could be effected by dilution in the case of cyclopentadienylidene [137] and dibenzocycloheptatrienylidene [138,139].

Loss of stereospecifity on dilution with cyclohexane also occurs in the otherwise stereospecific addition of the biscarbene *55'* by photolysis of *55* in cis-2-butene [139].

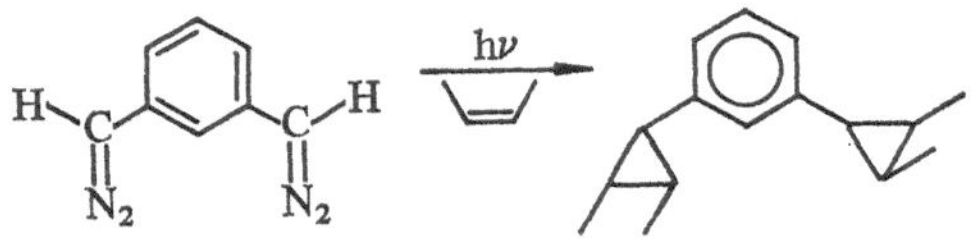

3.2.4. Triplet Carbenes by Sensitization

As has already been mentioned (vide supra, p. 118) the population of the triplet state of a carbene depends on the effectiveness of the intersystem crossing steps k_{isc} and k'_{isc}. These rates can be altered by dilution. Another technique which exclusively populates the triplet state of a carbene is *sensitization*, or *energy transfer*. A triplet sensitizer is required for this purpose — usually an aromatic ketone. In these ketones the intersystem crossing efficiency is almost 100%. Energy is then transferred from the sensitizer triplet to the diazoalkane, thus populating the triplet state of the latter.

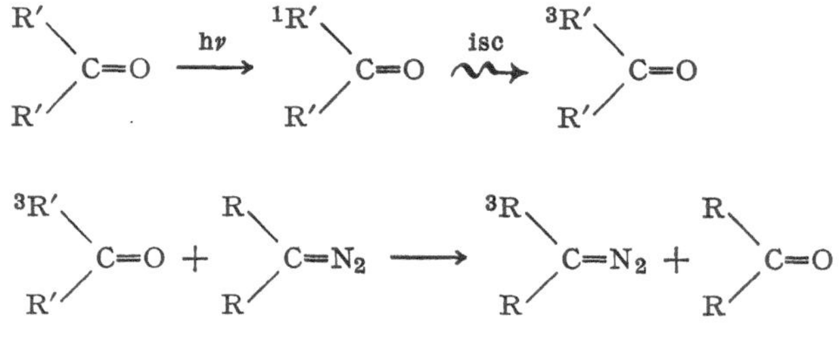

44

As shown by the scheme on p. 118, the triplet diazoalkene *44* can lose N_2 to give the triplet carbene *45*, which can now add to the olefin (b) in a non-stereospecific manner. Since the triplet state of the diazoalkane *44* has a longer lifetime, it can also add to the olefin directly, producing the biradical *47* which after extrusion of N_2 can undergo nonstereospecific ring closure to *43* and *46* (c). This possibility should also be kept in mind in interpreting the reactions of a triplet carbene generated by energy transfer. Care must also be taken to ensure a complete light capture by the sensitizer in the presence of the diazoalkane, which usually has a high extinction coefficient invariably extending into the visible region.

Thus for instance, in the sensitization experiments carried out with ethoxy-carbonyl-carbene, dimethoxy-carbonylcarbene, and carbene A, the light capture was generally more than 80% [135]. In the case of phenyl-cyclopentadienylidenes the nonstereospecificity clearly depends on the light capture by the sensitizer [140].

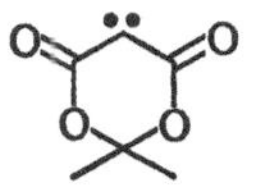

A

In the sensitized generation of ethoxy-carbonyl-carbene [141] the light capture by the diazocompound is only 25% [142]. A number of photosensitized reactions of triplet carbenes with cis- and trans-olefins are listed in Table 10. The sensitizers used, as well as their various triplet energies E_T, are also given in the table.

It is evident from Table 10 that identical mixtures of cis- and trans-cyclo-propanes obtained from either cis- or trans-olefins are quite rare. Only the cyclohexanone-carbene *59* [141] and methyl-bisalkoxy-carbonyl-carbene [123] give

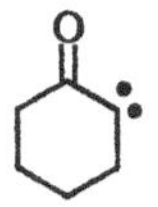

59

the same cis-/trans-ratio in the products. Ethyl-alkoxy-carbonyl-carbene also tends to give identical mixtures [125,142]. All other triplet-carbenes — generated by energy transfer — give different cis-/trans-cyclopropane ratios with different olefins. However, the increase of nonstereospecifity in the sensitized reactions clearly indicates that it is the triplet which is involved in these reactions.

In the photosensitized reaction of diazomethane which yields triplet methylene, a loss of stereospecifity is observed. However, with trans-2-butene cyclo-addition occurs only to a limited extent.

Photolysis of methyl-phenyl-diazomethane in cis-butene gives mainly styrene formed by hydride shift and the isomeric cycloaddition products *60—62* [139].

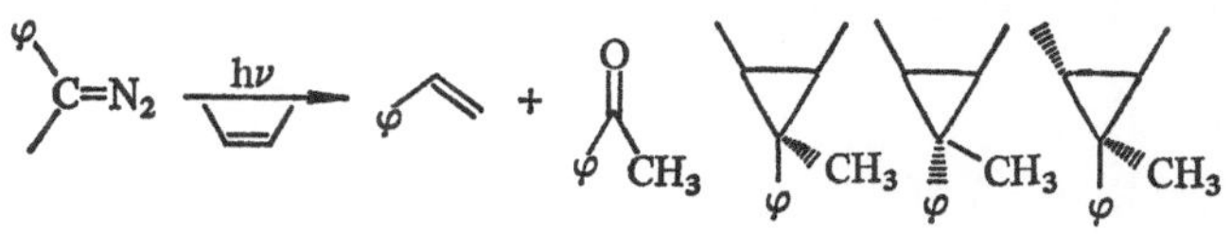

			$60 + 61 + 62$	$62/60 + 61$
			60 *61* *62*	
Direct:	6.0	2.3	6.9	4/96
Sensitized: (φ_2CO)	3.7	20.2	6.9	8/92

H. Dürr

Table 10. Photosensitized decomposition of diazo compounds in the presence of olefins

Diazoalkane (hν)	Sensitizer E_T in kcal/m	trans Cyclopropane in (%) from cis-olefin	cis Cyclopropane in (%) from trans-olefin	Ref.
H_2CN_2	—	0 [1]	0 [1]	
	Benzophenone (68.5)	34 [1]	Trace [1]	104)
$PhCN_2CH_3$	—	—	—	
	Benzophenone	9 [1]	4 [1]	
		8 [1]	—	139)
		15 [1]	—	
$PhCOCHN_2$	—	50 [1]	27 [1]	
	Michlers Ketone (61)	55 [1]	26 [1]	125)
CH_3COCHN_2	Benzophenone	56 [1]	11 [1]	141)
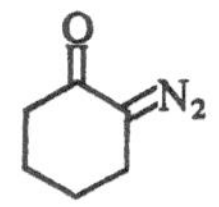	—	86 [1]	14 [1]	141)
$H_5C_2O_2CHCN_2$	—	0	0	135)
	Benzophenone	74 [1]	9 [1]	142)
		67 [1]	10 [1]	
$(CH_3)_3SiCN_2CO_2C_2H_5$	—	5 [1]	—	122)
	Benzophenone	17 [1]	—	
$(ROOC)_2C=N_2$	—	8 [1]	10 [1]	123)
	Benzophenone	90 [1]	14 [1]	
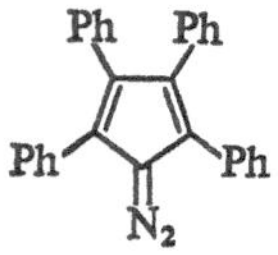	—	< 1 [2]	< 1 [2]	143)
	Xanthone (74)	27 [2,3]	<27 [2,3]	140)
	Benzophenone	5—10 [2]	5—10 [2]	
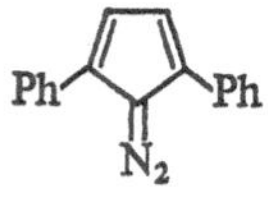	—	< 1 [2]	—	140)
	Benzophenone	5—10 [2,3]	—	

1) cis- and trans-But-2-ene.
2) cis- and trans-4-Methyl-pent-2-ene.
3) Light capture by the sensitizer was not complete (1:2) [140].

The photosensitized decomposition reduces the hydride shift, as may be expected for a triplet species. However, the stereochemical change in the cycloaddition is relatively small. Similiar results were obtained by Baer and Gutsche for the sensitized photolysis of o-n-butyl-phenyl-diazomethane 63 [144].

126

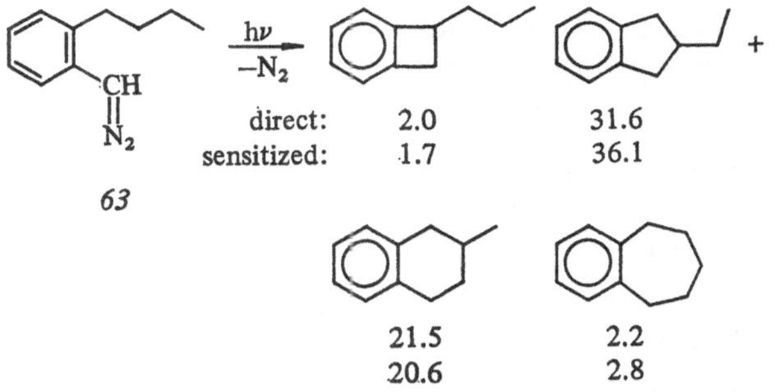

direct:	2.0	31.6
sensitized:	1.7	36.1

63

21.5	2.2
20.6	2.8

Here, too practically no change between the direct and the sensitized reaction can be observed.

Phenylcarbene generated by different reaction routes always gives a similar cycloaddition pattern with butenes [145]. These results indicate that an equilibrium is present between singlet and triplet phenylcarbene.

Photochemically generated benzoylcarbene (from diazo-aceto-phenone) shows little change in the stereochemistry of the cycloaddition as between direct and sensitized photolysis. This could also be attributed to a singlet-triplet-equilibrium [125].

Diazoacetone and diazocyclohexanone give cycloaddition only in the sensitized reaction, which yields the corresponding cyclopropanes 64—66 and 67, 68 [141].

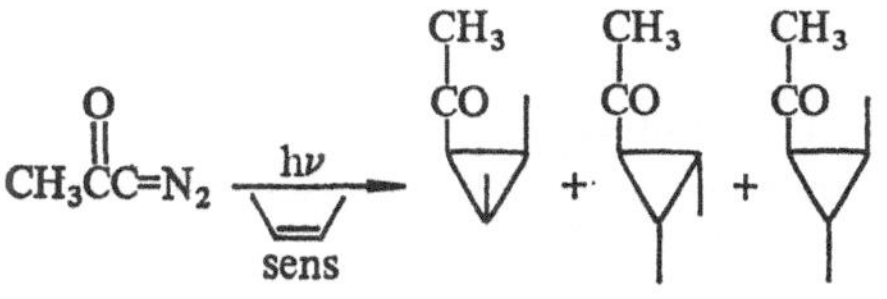

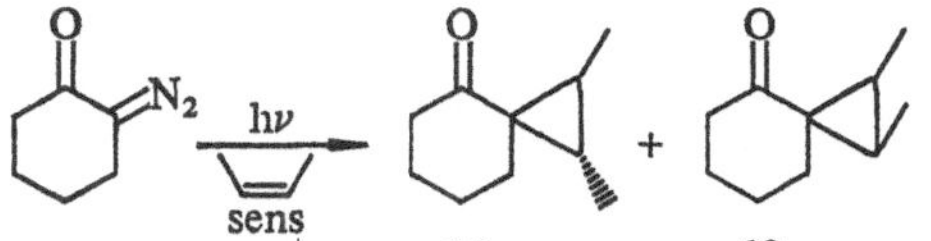

These reactions are only observed with the triplet carbene. The singlet carbene exclusively reacts to give a Wolff-rearrangement. Alkoxycarbonyl-carbene [135,142] as well as the carbena-cyclopentadienes [143,140] are exceptional since on direct photolysis of the corresponding diazoalkanes they add *stereospecifically*; however the sensitized reaction yields an appreciable amount of *nonstereospecifically* formed products due to the triplet. These data clearly preclude a singlet—triplet equilibrium for these carbenes. A purely triplet carbene is produced by energy transfer to diazo-malonate, which adds in inverse cis/trans-ratios in the direct and the sensitized reactions respectively. Here the triplet bis-carbomethoxy-carbene is very readily produced. Relative reactivities have also been determined for singlet and triplet biscarbomethoxy-carbene (see Table 11) [135].

Table 11. Relative rates of addition of singlet and triplet bis(methoxy carbonyl) carbene to olefins

	Singlet A	Triplet A
2.3-Dimethyl-2-butene	0.88	0.33
2-Methyl-2-butene	1.0	1.0
1-Pentene	0.47	0.46
3.3-Dimethyl-1-butene	0.48	0.48
cis-4-Methyl-2-pentene	0.55	0.15
trans-4-Methyl-2-pentene	0.23	0.13
2.3-Dimethyl-1.3-butadiene	1.3	4.4
1.3-Butadiene	—	4.5

It is evident from Table 11 that the rate of addition of the triplet bis(methoxy-carbonyl) carbene is somewhat slower than that of the singlet. Another important general rule may also be deduced from Table 11: the triplet carbene adds to dienes about 3—4 times faster than to olefins: the reactivity ratio of 2,3-dimethyl-buta-diene-1,3/pentene-1 is 9.6 for triplet and 2.8 for the singlet. This ratio may be compared with that for diphenylcarbene (1,3-butadiene/hexene-1), which is > 100.

Table 12. Relative addition rates of singlet and triplet 69 [24]

Olefin	About 100% T_1-69 (90% hexafluorobenzene)	About 50% T_1-69 (pure olefin)
	0.37	0.6
	1.00	1.00
Pentene-1)	1.70	0.47
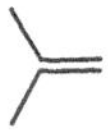	0.86	0.38
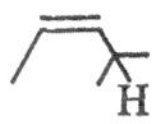	0.40	0.43
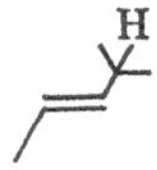	0.69	0.56
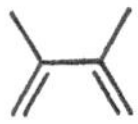	9.0	3.5

Similar relative data for the addition rates of singlet and triplet fluorenylidene 69 have been obtained using the dilution technique [24].

In the case of cyclopropyl-diazo-acetate 72 energy transfer reduces the extent of intramolecular in favour of the intermolecular reactions [146].

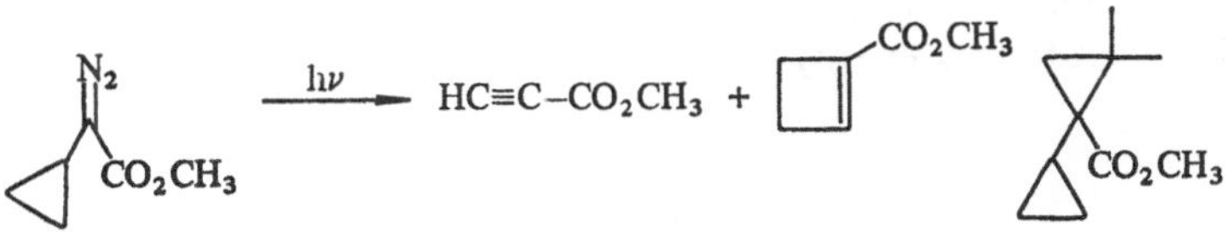

		72	73	74
	direct:	18%	67%	—
	sensitized:	16%	14%	—
in 	direct:	3%	20%	8,5%
	sensitized:	3%	11%	33

Dibenzo-cycloheptatrienylidene 70 shows no difference in behavior when generated in the presence of a sensitizer [138,139].

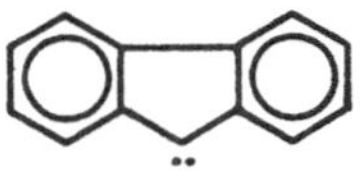

69

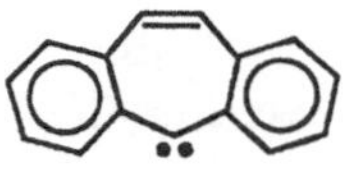

70

Different Reactions of Triplet and Singlet Carbenes

It would be interesting to find selective reactions of triplet carbenes not given by the corresponding singlet carbenes. In the reactions presented above a difference in stereochemistry or relative reactivity of triplet and singlet carbenes was the only one found. However, some reactions seem restricted to one spin state only; e.g., the triplet carbene gives products which differ from those given by the singlet reactant (see also p. 127).

An example are the diazoketones:

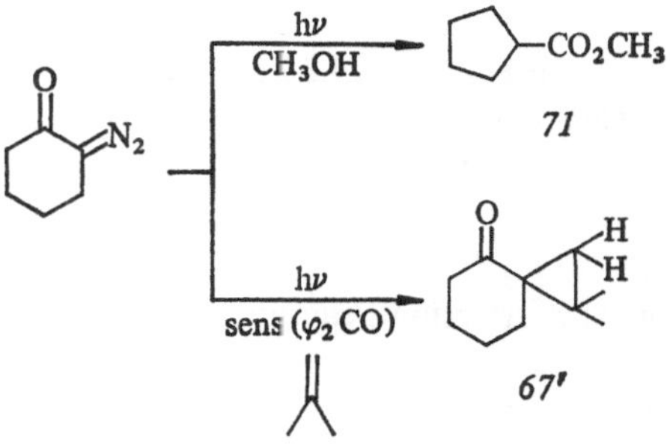

H. Dürr

The sensitized reaction yields the cyclopropanes *67'* as the only product, whereas direct photolysis results in a Wolff-rearrangement yielding the cyclopentaneester *71* [141]. A Wolff-rearrangement is only given by the singlet carbene [147], while the triplet gives addition or hydrogen abstraction [125]. Bis-methoxy-carbonyl-carbene reacts with allyl compounds to give insertion products by an ylide mechanism. However, the "insertion reaction" is suppressed if the triplet carbene is generated by energy transfer.

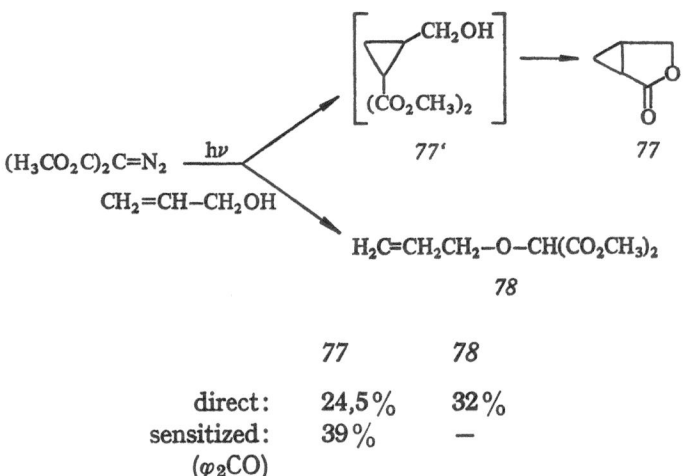

	75	*76*
direct:	23%	53%
Sens (φ_2CO):	88%	5%

	77	*78*
direct:	24,5%	32%
sensitized: (φ_2CO)	39%	—

The intermediate in the insertion reaction is an ylide which reacts only via singlet carbene [106, 148].

the Only cycloaddition is also given by the triplet carbene *80*.

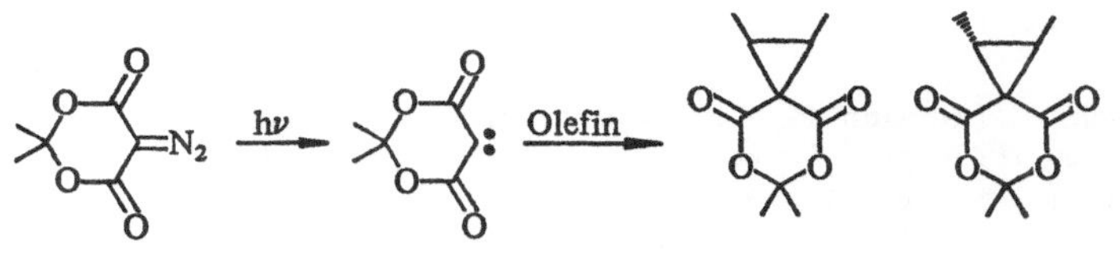

79   80   81   82

cis-2-butene: direct: — —
      sens. (φ_2CO): 14 86

trans-2-butene: direct: — —
      sens.: 15 85

These data demonstrate that a triplet carbene 80 is produced, which gives the same ratio of products 81/82 with the cis- and with the trans-olefin [135]. A clear differentiation of the products formed is observed in the reaction of fluorenylidene 69 with dicyclopropylethylene.

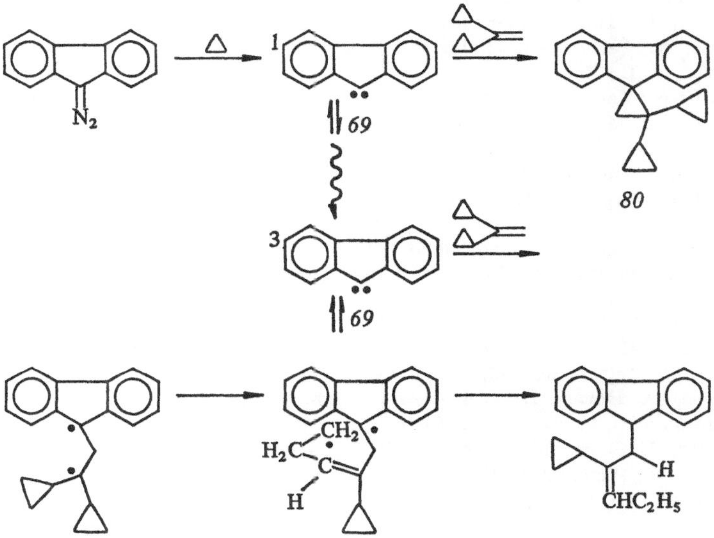

81   82   83

ratio: 83/80
Neat: 0.1
30 times dil. with decalin: 3.0

On dilution with decalin the original ratio 83/80 increases from 0.1 to 3.0, which is a clear indication that 83 is formed from the triplet and 80 from the singlet fluorenylidene [149].

Der Deutschen Forschungsgemeinschaft sowie dem Fonds der Chemischen Industrie sei an dieser Stelle für die Förderung der hier zitierten eigenen Arbeiten gedankt.

131

H. Dürr

4. References

1) Hantzsch, A., Lehmann, M.: Ber. Deut. Chem. Ges. *34*, 2522 (1901).
2) Staudinger, H., Kupfer, O.: Ber. Deut. Chem. Ges. *45*, 501 (1912).
3) Hine, J.: J. Am. Chem. Soc. *72*, 2438 (1950).
4) Doering, W. v. E., Hoffmann, A. K.: J. Am. Chem. Soc. *76*, 6162 (1954).
5) Kirmse, W.: Angew. Chem. *71*, 537 (1959); *73*, 161 (1961).
6) Miginiac, P.: Bull. Soc. Chim. France p. 2000 (1962).
7) Chinoporos, E.: Chem. Rev. *63*, 235 (1963).
8) Kirmse, W.: Carbene chemistry. New York: Academic Press 1964.
9) Cadogan, J. I. G., Perkins, M. J.: The chemistry of alkenes (S. Patai, ed.), p. 633. New York: Wiley 1964.
10) De More, W. B., Benson, S. W.: Advan. Photochem. *2*, 219 (1964).
11) Frey, H. M.: Progr. Reaction Kinetics 2, 131 (1964).
12) Hine, J.: Divalent carbon. New York: Ronald Press 1964.
13) Bell, J. A.: Progr. Phys. Org. Chem. 2, 1 (1964).
14) Parham, W. E., Schweizer, E. E.: Org. Reactions *13*, 55 (1964).
15) Rees, C. W., Smithen, C. E.: Advan. Heterocycl. Chem. *3*, 57 (1964).
16) Kirmse, W.: Angew. Chem. *77*, 1 (1965); Angew. Chem. Intern. Ed. Engl. *4*, 1 (1965).
17) Herold, B. J., Gaspar, P. P.: Topics Curr. Chem. *5*, 89 (1965).
18a) Köbrich, G.: Angew. Chem. *79*, 15 (1967); Angew. Chem. Intern. Ed. Engl. *6*, 41 (1967);
   b) Closs. G. L.: Topics Stereochem. *3*, 193 (1968).
19) Kirmse, W.: Carbene carbenoide and carben-analoge. Weinheim: Verlag Chemie 1969.
20) Gilchrist, T. L., Rees, C. W.: Carbenes, nitrenes and arynes. London: Nelson 1969.
21) Moss, R. A.: Chem. Eng. News *47*, 30 (1969).
22) v. Leusen, A. M., Strating, J.: Quart Rep. Sulfur Chem. *5*, 67 (1970).
23) Kirmse, W.: Carbene chemistry, New York: Academic Press 1971.
24) Jones, M. jr., Moss, R. A.: Carbenes. New York: John Wiley and Sons 1973.
25) Bethell, D.: Organ. react. intermediates (S. P. Manns, ed.) New York: Academic Press, 1973.
26) Dürr, H.: Topics Curr. Chem. *40*, 103 (1973).
27) Dürr, H.: Photochemie, Vol. IV/5, Houben-Weyl. Stuttgart: Georg Thieme Verlag, in print.
28) Hoffmann, R.: Tetrahedron *22*, 539 (1966).
29a) Brinton, R. K., Vollman, D. H.: J. Chem. Phys. *19*, 1394 (1951); see however: Csizmadia J. G., *et al.*: Tetrahedron *25*, 2121 (1969);
   b) Kirmse, W., Horner, L.: Liebigs Ann. Chem. *625*, 34 (1959).
30) Heilbronner, E., Martin, H. D.: Chem. Ber. *106*, 3376 (1973).
31) Herzberg, G., Shoosmith, J.: Nature 1801 (1959).
32) Moore, C. B., Pimentel, G. C.: J. Chem. Phys. *41*, 3504 (1964).
33) Borodko, Y. G., Shilov, A. E., Shteinmann, A. A.: Dokl. Acad. Nauk SSSR *168*, 581 (1966).
34) Closs, G. L., Böll, W. A., Heyn, H., Dev, V.: J. Am. Chem. Soc. *90*, 173 (1968).
35) Franck-Neumann, M., Buchecker, C.: Tetrahedron Letters 1969, 15.
36) Dürr, H., Sergio, R., Gombler, W.: Angew. Chem. *84*, 215 (1972).
37) Kirmse, W.: Chem. Ber. *93*, 2357 (1960).
38) Dolbier, W. R., Williams, W. M.: J. Am. Chem. Soc. *91*, 2818 (1969).
39) Scheiner, P.: J. Org. Chem. *34*, 199 (1969).
40) Fuchs, B., Rosenblum, M.: J. Am. Chem. Soc. *90*, 1061 (1968).
41) Shantarovitch, P. S.: Dokl. Acad. Nauk SSSR *116*, 255 (1957).
42) Setser, D. W., Rabinovitch, B. S.: Can. J. Chem. *40*, 1425 (1962).
43) Dunning, W. J., McCain, C. C.: J. Chem. Soc. B *1966*, 68.
44) Murgulscu, G., Onescu, T.: J. Chim. Phys. Physicochim. Biol. *58*, 508 (1961).
45) Reimlinger, H.: Chem. Ber. *97*, 339 (1964).
46) Bethell, D., Whittaker, D.: J. Chem. Soc. B *1966*, 778.
47) Forster, J. M., Boys, S. F.: Rev. Mod. Phys. *32*, 305 (1960).
48) Harrison, J. F., Allen, L. C.: J. Am. Chem. Soc. *91*, 807 (1969).

[49] Harrison, J. F., Allen, L. C.: J. Am. Chem. Soc. *93*, 4112 (1971).
[50] Hoffmann, R., Zeiss, G. D., van Dine, G. W.: J. Am. Chem. Soc. *90*, 1485 (1968).
[51] Bender, C. F., Schaefer, H. F.: J. Am. Chem. Soc. *92*, 4984 (1970). — O'Neil, S. V., Schaefer, H. F., Bender, C. F.: J. Chem. Phys. *55*, 162 (1971).
[52a] Bodor, N., Dewar, M. J. S., Wasson, J. S.: J. Am. Chem. Soc. *94*, 9095 (1972).
[b] Dewar, M. J. S., Haddon, R. C., Weiner, P. K.: J. Am. Chem. Soc. *96*, 254 (1974).
[53] Lathan, W. A., Hehre, W. J., Pople, J. A.: J. Am. Chem. Soc. *93*, 808 (1971). — Lathan, W. A., Hehre, W. J., Curtiss, L. A., Pople, J. A.: J. Am. Chem. Soc. *93*, 6377 (1971). — Chupka, W. A., Lifshitz, C.: J. Chem. Phys. *48*, 1109 (1968). — Chupka, W. A.: J. Chem. Phys. *48*, 2337 (1968).
[54] Hase, W. L., Philips, R. J., Simons, J W.: Chem. Phys. Letters *12*, 161 (1971).
[55] Frey, H. M.: J. Chem. Soc. Chem. Commun. *1972*, 1024.
[56] Gleiter, R., Hoffmann, R.: J. Am. Chem. Soc. *90*, 5457(1968).
[56a] Salem, L., Stohrer, W. D.: unpublished results.
[57] Skell, P. S., Cholod, M. S.: J. Am. Chem. Soc. *91*, 7131 (1969).
[58] Dürr, H., Werndorff, F.: Angew. Chem. *86*, 413 (1974).
[59] Christensen, L. W., Waali, E. E., Jones, W. M.: J. Am. Chem. Soc. *94*, 2118 (1972).
[60] Wassermann, E., Yager, W. A., Kuck, V. J.: Chem. Phys. Letters *7*, 409 (1970). — Wassermann, E., Kuck, V. J., Hutton, R. S., Yager, W. A.: J. Am. Chem. Soc. *92*, 7491 (1970).
[61] Bernheim, R. A., Bernhard, H. W., Wang, P. S., Wood, L. S., Skell, P. S.: J. Chem. Phys. *54*, 3223 (1971).
[62] Wassermann, E., Barash, L., Yager, W. A.: J. Am. Chem. Soc. *87*, 4974 (1965).
[63] Murray, R. W., Trozzolo, A. M., Wasserman, E., Yager, W. A.: J. Am. Chem. Soc. *84*, 3213, 4990 (1962).
[64] Trozzolo, A. M., Wasserman, E., Yager, W. A.: J. Am. Chem. Soc. *87*, 129 (1965).
[65] Wasserman, E., Trozzolo, A. M., Yager, W. A., Murray, R. W.: J. Chem. Phys. *40*, 2408 (1964).
[66] Hutchinson, C. A., jr., Kohler, B.: J. Chem. Phys. *51*, 3327 (1969).
[67] Bernheim, R. A., Kempf, R. J., Gramas, J. V., Skell, P. S.: J. Chem. Phys. *43*, 196 (1965).
[68] Bernheim, R. A., Kempf, R. J., Humer, P. W., Skell, P. S.: J. Chem. Phys. *41*, 1156 (1964).
[69] Wasserman, E., Barash, L., Trozzolo, A. M., Murray, R. W., Yager, W. A.: J. Am. Chem. Soc. *86*, 2304 (1964).
[70] Murahashi, S., Moritani, I., Nagai, T.: Bull. Soc. Japan *40*, 1655 (1967).
[71] Moritani, I., Murahashi, S. I., Yamamoto, M. H., Itoh, K., Mataga, N.: J. Am. Chem. Soc. *89*, 1259 (1967). — Brandon, R. W., Closs, G. L., Hutchison, C. A.: J. Chem. Phys. *37*, 1878 (1962).
[72] Wasserman, E., Murray, R. W.: J. Am. Chem. Soc. *86*, 4203 (1964).
[73] Jones, M., jr., Harrison, A. M., Retting, K. R.: J. Am. Chem. Soc. *91*, 7462 (1969).
[74] Moriconi, E. J., Murray, J. J.: J. Org. Chem. *29*, 3577 (1964).
[75] Trozzolo, A. M., Murray, R. W., Smolinsky, G., Yager, W. A., Wasserman, E.: J. Am. Chem. Soc. *85*, 2526 (1963); *89*, 5076 (1967).
[76] Herzberg, G.: Proc. Roy. Soc. (London) Ser. A *262*, 291 (1961).
[77] Herzberg, G., Johns, J. W.: J. Chem. Phys. *54*, 2276 (1971).
[78] De More, W. B., Pritchard, H. O., Davidson, N.: J. Am. Chem. Soc. *81*, 5874 (1959).
[79] Gibbons, W. A., Trozzolo, A. M.: J. Am. Chem. Soc. *88*, 172 (1966). — Trozzolo, A. M., Gibbons, W. A.: J. Am. Chem. Soc. *89*. 239 (1967).
[80] Baird, M. S., Dunkin, I. R., Poliakoff, M.: Contr. Pap. No. 14, IUPAC-Symposium, Enschede 1974.
[81] Moritani, I., Murahashi, S. I., Ashitaka, H., Kimura, K., Tsubomura, H.: J. Am. Chem. Soc. *90*, 5918 (1968).
[82] Yamamoto, Y., Moratini, I., Maeda, Y., Murahashi, S.: Tetrahedron *26*, 251 (1970). — Moratini, I., Murahashi, S. I., Nishinc, M., Kimura, K., Tsubomura, H.: Tetrahedron Letters *1966*, 373.

133

[83] Lepley, A. R., Closs, G. L.: Chem. induced magnetic polarization. New York: John Wiley and Sons 1973.

[84] Closs, G. L.: J. Am. Chem. Soc. *91*, 4552 (1969). — Closs, G. L., Trifunac, A. D.: J. Am. Chem. Soc. *92*, 2186 (1970).

[85] Kaptein, R.: J. Chem. Soc. Chem. Commun. *1971*, 732.

[86] Closs, G. L., Closs, L. E.: J. Am. Chem. Soc. *91*, 4549, 4550 (1969).

[87] Closs, G. L., Trifunac, A. D.: J. Am. Chem. Soc. *92*, 7227 (1970).

[88] Kaplan, M. L., Roth, H. D.: J. Chem. Soc. Chem. Commun. *1972*, 970.

[89] Roth, H. D.: J. Am. Chem. Soc. *94*, 1761 (1972).

[90] Roth, H. D.: J. Am. Chem. Soc. *93*, 1527 (1971); Padwa, A., Eisenhardt, W.: J. Am. Chem. Soc. *93*, 1400 (1972).

[91] Cocivera, M., Roth, H. D.: J. Am. Chem. Soc. *92*, 2573 (1970).

[92] Iwamura, H. Imahashi, Y., Kushida, K.: J. Am. Chem. Soc. *96*, 921 (1974).

[93] Dobson, R. C., Hayes, D. M., Hoffmann, R.: J. Am. Chem. Soc. *93*, 6188 (1971).

[94] Ring, D. F., Rabinovitch, B. S.: J. Am. Chem. Soc. *88*, 4285 (1965).

[95] Ring, D. F., Rabinovitch, B. S.: Can. J. Chem. *46*, 2435 (1968).

[96] Doering, W. v. E., Prinzbach, H.: Tetrahedron *6*, 24 (1959).

[97] Herzog, B. M., Carr, R. W.: J. Phys. Chem. *71*, 2688 (1967).

[98] Frey, H. M., Voisey, M. A.: Trans. Faraday Soc. *64*, 961 (1968). — Voisey, M. A.: Trans. Faraday Soc. *64*, 3058 (1968).

[99] Bethell, D., Whittaker, D., Callister, J. D.: J. Chem. Soc. *1965*, 2466.

[100] Nozaki, H., Nakano, M., Kondo, K.: Tetrahedron *22*, 477 (1966).

[101] Kirmse, W., Horner, L., Hoffmann, H.: Liebigs Ann. Chem. *614*, 19 (1958).

[102] Cauquis, G., Reverdy, G.: Tetrahedron Letters *1967*, 1493.

[103] Fleming, J. C., Shechter, H.: J. Org. Chem. *34*, 3962 (1969).

[104] Kopecky, K. R., Hammond, G. S., Leermakers, P. A.: J. Am. Chem. Soc. *83*, 2397 (1961); *84*, 1015 (1962).

[105] Jones, M., jr., Ando, W., Kulczycki, A.: Tetrahedron Letters *1967*, 1391.

[106] Ando, W., Imai, I., Migita, T.: J. Org. Chem. *37*, 3596 (1972).

[107] Kaufman, J. A., Weininger, S. J.: J. Chem. Soc. Chem. Commun. *1969*, 593.

[108] Skell, P. S., Plonka, J. H.: J. Am. Chem. Soc. *92*, 836, 5620 (1970). — Skell, P. S., Engel, R. R.: J. Am. Chem. Soc. *88*, 3749, 4883 (1966).

[109] Hoffmann, R.: J. Am. Chem. Soc. *90*, 1475 (1968).

[110] Skell, P. S., Woodworth, R. C.: J. Am. Chem. Soc. *78*, 4496 (1956); *81*, 3383 (1959).

[111] Frey, H. M.: Proc. Roy. Soc. (London) Ser. A *251*, 409, 575 (1959).

[111a] Halberstadt, M. L., Mc Nesby, J. R.: J. Am. Chem. Soc. *89*, 3417 (1967).

[112] Simons, J. W., Taylor, G. W.: J. Phys. Chem. *73*, 1274 (1969).

[113] Bader, R. F., Generosa, J. I.: Can. J. Chem. *43*, 1631 (1965).

[114] Rabinovitch, B. S., Watkins, K. W., Ring, D. F.: J. Am. Chem. Soc. *87*, 4960 (1965).

[115] Frey, H. M.: J. Chem. Soc. Chem. Commun. *1965*, 260.

[116] Elliot, C. S., Frey, H. M.: Trans. Faraday Soc. *64*, 2352 (1968).

[117] Eder, T. W., Carr, R. W. jr.: Phys. Chem. *73*, 2074 (1969).

[118] Atherton, J. H., Fields, R.: J. Chem. Soc. (C) *1967*, 1450.

[119] Skell, P. S., Klebe, J.: J. Am. Chem. Soc. *82*, 247 (1960).

[120] Closs, G., Moss, R. A.: J. Am. Chem. Soc. *86*, 4042 (1964). — Closs, G. L., Closs, L. E.: Angew. Chem. *74*, 431 (1962); Similar results showing even less stereospecifity were obtained for nitro-phenylcarbenes: Closs, G. L., Goh, S.: J. Chem. Soc. Perkin Trans. I *1972*, 2103. — Goh, S. H.: J. Chem. Soc. C *1971*, 2275.

[121] Baron, W. J., Hendrick, M. E., Jones, M., jr.: J. Am. Chem. Soc. *95*, 6286 (1973).

[122] Schöllkopf, U., Hoppe, D., Rieber, N., Jacobi, V.: Liebigs Ann. Chem. *730*, 1 (1969).

[123] Jones, M., jr., Kulczycki jr., A., Hummel, K. F.: Tetrahedron Letters *1967*, 183.

[124] Ciganek, E.: J. Am. Chem. Soc. *88*, 1979 (1966).

[125] Cowan, D. O., Couch, M. M., Kopecky, K. R., Hammond, G. S.: J. Org. Chem. *29*, 1922 (1964).

[126] Mc Bee, E. T., Bosoms, J. A., Morton, C. J.: J. Org. Chem. *31*, 768 (1966).

[127] Mc Bee, E. T., Sienkowski, J. K.: J. Org. Chem. *38*, 1340 (1973).

[128] Jones, M., jr., Rettig, K. R.: J. Am. Chem. Soc. *87*. 4013 (1965).

129) Koser, G. F., Pirkle, W. H.: J. Org. Chem. *32*, 1992 (1967).
130) Cauquis, G., Reverdy, G.: Tetrahedron Letters *1968*, 1085, 3771.
131) Cauquis, G., Reverdy, G.: Tetrahedron Letters *1971*, 4289.
132) Cauquis, G., Reverdy, G.: Tetrahedron Letters *1972*, 3491.
133) Moss, R. A., Dolling, U. H.: J. Am. Chem. Soc. *93*, 954 (1971).
134) Ring, D. F., Rabinovitch, B. S.: J. Phys. Chem. *72*, 191 (1968).
135) Jones, M., jr., Ando, W., Hendrick, M. E., Kulczycki jr., A., Howely, P. M., Hummel, K. F., Malament, D. S.: J. Am. Chem. Soc. *94*, 7469 (1972).
136) Pirkle, W. H., Koser, G. F.: Tetrahedron Letters *1968*, 3959.
137) Moss, R. A., Przybyla, J. R.: J. Org. Chem. *33*, 3816 (1968).
138) Murahashi, S. I., Moritani, I., Nishino, M.: Tetrahedron *27*, 5131 (1971).
139) Murahashi, S. I., Yoshimura, Y., Yamamoto, Y., Mortani, I.: Tetrahedron *28*, 1485 (1972). — Moritani, I., Yamamoto, Y., Murahashi, S. I.: Tetrahedron Letters *1968*, 5697, 5755.
140) Dürr, H., Bujnoch, W.: Tetrahedron Letters *1973*, 1433.
141) Jones, M., jr., Ando, W.: J. Am. Chem. Soc. *90*, 2200 (1968). — Jones, M., jr., Wataru, A.: J. Am. Chem. Soc. *90*, 2200 (1968).
142) Reetz, M., Schöllkopf, U., Banhidai, B.: Liebigs Ann. Chem. *1973*, 599.
143) Dürr, H., Scheppers, G.: Chem. Ber. *100*, 3236 (1967).
144) Baer, T. A., and Gutsche, C. D.: J. Am. Chem. Soc. *93*, 5180 (1971).
145) Scheiner, P.: J. Org. Chem. *34*, 199 (1969); Tetrahedron Letters *1971*, 4489. — Dietrich, H., Griffin, G. W., Petterson, R. C,: Tetrahedron Letters *1968*, 153.
146) Sohn, M. B., Jones, M., jr.: J. Am. Chem. Soc. *94*, 8280 (1972).
147) Padwa, A., Layton, R.: Tetrahedron Letters *1965*, 2167.
148) Ando, W., Kondo, S., Migita, T.: J. Am. Chem. Soc. *91*, 6516 (1969); Bull. Soc. Chem. Japan *44*, 571 (1971).
149) Shimizu, N., Nishida, S.: J. Chem. Soc. Chem. Commun. *1972*, 389.

Received September 30, 1974

H. Meier
Die Photochemie
der organischen Farbstoffe

168 Abbildungen. XVI, 471 Seiten
1963 (Organische Chemie in
Einzeldarstellungen, Band 7)
DM 98,—; US $ 40.20
ISBN 3-540-03034-4

Inhaltsübersicht: Die Lichtab-
sorption der Farbstoffe.
Lumineszenz der Farbstoffe. Die
photochemischen Umsetzungen
an organischen Farbstoffen. Der
lichtelektrische Effekt der organischen Farbstoffe. — Spezielle
Reaktionen: Die spektrale Sensi-
bilisierung der photographischen
Schicht. Die Farbstoff-Sensibili-
sierung des Photoeffekts anorga-
nischer Halbleiter. Der photody-
namische Effekt. Der Sehvorgang.
Die Photosynthese. — Anhang:
Zum Problem der Energieüber-
tragung.

A. Schönberg
**Preparative
Organic Photochemistry**

In cooperation with G.O. Schenck,
O.A. Neumüller. Second, com-
pletely revised edition of Präpa-
rative Organische Photochemie.
4 figures and 51 tables
XXIV, 608 pages. 1968
Cloth DM 165,—; US $67.70
ISBN 3-540-04325-X

This monograph is an exhaustive
compilation of photochemical
reactions that are of interest to
the organic chemist. Sufficient
experimental details are given to
make this book a highly useful
manual of photochemical laboratory techniques for the student
and the specialist.

O. Stasiw
Elektronen- und Ionenprozesse
in Ionenkristallen
mit Berücksichtigung
photochemischer Prozesse

107 Abbildungen. VIII, 307 Seiten
1959 (Struktur und Eigenschaften
der Materie in Einzeldarstellungen,
Band 22)
Gebunden DM 80,—; US $32.80
ISBN 3-540-02475-1

Inhaltsübersicht: Statistik von
Störstellen in Ionenkristallen. —
Fehlerordnungsenergie. — Platz-
wechselvorgänge, Diffusion und
Ionenleitung. — Das Absorptions-
spektrum des idealen Ionengitters.
— Das Absorptionsspektrum von
Ionengittern mit stöchiometrischem
Überschuß der Kationen oder
Anionenkomponente. — Absorptionsspektren von Ionengittern mit
Fremdzusätzen. — Elektronische
Störstellentheorie. — Halbleiter-
prozesse. — Lichtelektrische Lei-
tung. — Photochemische Prozesse
in reinen Ionengittern. — Photochemische Prozesse in Ionengittern
mit Zusätzen. — Photochemische
Prozesse in mechanisch verformten
Kristallen. — Störstellen und Kern-
resonanz. — Anwendung der adia-
batischen Näherung auf Kristalle
mit Störstellen.

Preisänderungen vorbehalten
Prices are subject to change
without notice

**Springer-Verlag
Berlin   Heidelberg   New York**

Photochemie

6 Abbildungen. 197 Seiten (31 Seiten in Englisch). 1967
(Topics in Current Chemistry / Fortschritte der chemischen Forschung,
Band 7, Heft 3). DM 72,–: US $29.60 ISBN 3-540-03796-9

Inhaltsübersicht: R. Steinmetz: Photochemische Carbocyclo-Additions-
reaktionen. D. Elad: Some Aspects of Photoalkylation Reactions.
M. Pape: Die Photooximierung gesättigter Kohlenwasserstoffe.
E. Fischer: Photochromie und reversible Photoisomerisierung.

Photochemistry

11 figures. 224 pages. 1969 (Topics in Current Chemistry / Fortschritte der chemischen Forschung, Band 13, Heft 2)
DM 65,–: US $26.70 ISBN 3-540-04489-2

From the Contents: J.L.R. Williams: Photochemical Reactions of
Polymers. M.B. Rubin: Photochemistry of o-Quinones and -Diketones.
L.B. Jones, V.K. Jones: Photochemical Reactions of Cycloheptartrienes.
C. v. Sonntag: Strahlenchemie von Alkoholen.
E. Koerner von Gustorf, F.-W. Grevels: Photochemistry of Metal
Carbonyls, Metallocenes, and Olefin Complexes.

Photochemistry

50 figures. IV, 236 pages. 1974 (Topics in Current Chemistry /
Fortschritte der chemischen Forschung, Volume 46)
Cloth DM 68,–: US $27.90 ISBN 3-540-06592-X

From the Contents: J. Michl: Physical Basis of Qualitative MO Arguments in Organic Photochemistry. K.-D. Gundermann: Recent
Advances in Research on the Chemiluminescence of Organic Compounds. W.C. Herndon: Substituent Effects in Photochemical Cyclo-
addition Reactions. W.-D. Stohrer, P. Jacobs, K.H. Kaiser, G. Wiech,
G. Quinkert: Das sonderbare Verhalten elektronen-angeregter 4-Ring-
Ketone.

Preisänderungen vorbehalten
Prices are subject to change without notice

 **Springer-Verlag
Berlin Heidelberg New York**